p. 40, 43, 45, 47, 54

Black Holes:
A Bibliography with Indexes

BLACK HOLES:
A BIBLIOGRAPHY WITH INDEXES

LAWRENCE A. JAMESON (EDITOR)

Nova Science Publishers, Inc.
New York

Senior Editors: Susan Boriotti and Donna Dennis
Coordinating Editor: Tatiana Shohov
Office Manager: Annette Hellinger
Graphics: Wanda Serrano
Editorial Production: Jennifer Vogt, Matthew Kozlowski, Jonathan Rose, Alexandra Columbus and Maya Columbus
Circulation: Ave Maria Gonzalez, Vera Popovich, Luis Aviles, Melissa Diaz, Nicolas Miro and Jeannie Pappas
Communications and Acquisitions: Serge P. Shohov
Marketing: Cathy DeGregory

Library of Congress Cataloging-in-Publication Data
Available Upon Request

ISBN: 1-59033-287-3.

Copyright © 2002 by Nova Science Publishers, Inc.
    400 Oser Ave, Suite 1600
    Hauppauge, New York 11788-3619
    Tele. 631-231-7269   Fax 631-231-8175
    e-mail: Novascience@earthlink.net
    Web Site: http://www.novapublishers.com

All rights reserved. No part of this book may be reproduced, stored in a retrieval system or transmitted in any form or by any means: electronic, electrostatic, magnetic, tape, mechanical photocopying, recording or otherwise without permission from the publishers.

The authors and publisher have taken care in preparation of this book, but make no expressed or implied warranty of any kind and assume no responsibility for any errors or omissions. No liability is assumed for incidental or consequential damages in connection with or arising out of information contained in this book.

This publication is designed to provide accurate and authoritative information with regard to the subject matter covered herein. It is sold with the clear understanding that the publisher is not engaged in rendering legal or any other professional services. If legal or any other expert assistance is required, the services of a competent person should be sought. FROM A DECLARATION OF PARTICIPANTS JOINTLY ADOPTED BY A COMMITTEE OF THE AMERICAN BAR ASSOCIATION AND A COMMITTEE OF PUBLISHERS.

Printed in the United States of America

Contents

Preface   vii

Title Index   ix

Guide to Books   1

Journals and Magazines   13

Author Index   115

Subject Index   123

PREFACE

Black Holes are extremely dense celestial bodies that have been theorized to exist in the universe. They are the evolutionary endpoints of stars at least 10 to 15 times as massive as the Sun. If a star that massive or larger undergoes a supernova explosion, it may leave behind a fairly massive burned out stellar remnant. With no outward forces to oppose gravitational forces, the remnant will collapse in on itself. The star eventually collapses to the point of zero volume and infinite density, creating what is known as a "singularity". As the density increases, the path of light rays emitted from the star are bent and eventually wrapped irrevocably around the star. Any emitted photons are trapped into an orbit by the intense gravitational field; they will never leave it. Because no light escapes after the star reaches this infinite density, it is called a black hole. This bibliography sheds light on black holes by presenting the book literature especially but also some selected journal literature which is made accessible through the use of subject, author and title indexes.

TITLE INDEX

#

2D anti-de Sitter gravity as a conformally invariant mechanical system, 31
2D anti–de Sitter gravity as a conformally invariant mechanical system, 32
3D Grazing Collision of Two Black Holes, 14

A

A dying universe: the long-term fate and evolution of astrophysical objects, 13
A Fundamental Relation between Supermassive Black Holes and Their Host Galaxies, 50
A Global Uniqueness Theorem for Stationary Black Holes, 48
A heuristic way of obtaining the Kerr metric, 48
A model of a spheroidal body, 26
A NICMOS imaging study of high-z quasar host galaxies, 74
A physical model for the hard X-ray background, 109
A science odyssey--short trip. Blips, black holes, and the Nobel Prize., 1
A theory of thin shells with orbiting constituents, 23
Absolute conservation law for black holes, 57
Absorption of photons and fermions by black holes in four dimensions, 58
Accretion Rates onto Massive Black Holes in Four Quiescent Elliptical Galaxies, 110
Acoustic black holes: horizons, ergospheres and Hawking radiation, 107
Action and entropy of black holes in spacetimes with a cosmological constant, 33
AdS-CFT correspondence and the information paradox, 79
An action for black hole membranes, 90
An application of the Kerr black hole fly-wheel model to statistical properties of QSOs/AGNs, 88
An approximation for the rp-process, 94
An exact solution of the double-Kerr equilibrium problem, 82
An extreme critical spacetime: echoing and black-hole perturbations, 60
An observational constraint on gravitational lensing by objects of mass 109.5-1010.9 M, 16
Angelmass, 12
Angular momentum ambiguities in asymptotically flat spacetimes which are perturbations of stationary spacetimes, 104
Anti-de Sitter space and black holes, 20
Anti-de Sitter Space, Thermal Phase Transitionand Confinement in Gauge Theories, 110
Approximate analytical solutions to the initial data problem of black hole binary systems, 83
Approximate binary-black-hole metric, 15
Are higher order membranes stable in black hole spacetimes, 74
Are higher order membranes stable in black hole spacetimes?, 74
Area law corrections from state counting and supergravity, 34
As she climbed across the table, 5
Astrophysical evidence for the existence of black holes, 36
Astrophysics, 51
Asymmetric merger of black holes, 65
Atom made from charged elementary black hole, 51
Axially symmetric monopoles and black holes in Einstein-Yang-Mills-Higgs theory, 60

B

Bare masses in time-symmetric initial-value solutions for two black holes, 66

Bekenstein bound, holography and brane cosmology in charged black hole backgrounds, 33

Bertotti-Robinson-type solutions to dilaton-axion gravity, 39

Binaries for LISA, 106

Binary Black Hole Mergers from Planet-like Migrations, 57

Binary black-hole problem at the third post-Newtonian approximation in the orbital motion: Static part, 66

Black diamonds at brane junctions, 36, 37

Black diholes with unbalanced magnetic charges, 77

Black diholes, 48, 77

Black hole constraints on the running-mass inflation model, 75

Black hole demographics from the &formmu1, 85

Black Hole Entropy by the Brick-Wall Method in Four and FiveDimensions with U(1) Charges, 13

Black hole entropy calculations based on symmetries, 46

Black Hole Entropy from Conformal Field Theory in Any Dimension, 34

Black hole entropy reveals a twelfth dimension, 21

Black hole entropy: a spacetime foam approach, 97

Black hole evaporation and compact extra dimensions, 35

Black hole evolution by spectral methods, 71

Black hole excision for dynamic black holes, 14

Black hole formation from massive scalar fields, 56

Black hole formation in the Friedmann universe: Formulation and computation in numerical relativity, 98

Black hole gravitohydromagnetics, 6

Black hole masses from power density spectra: determinations and consequences, 43

Black hole physics: basic concepts and new developments, 3

Black hole polarization and new entropy bounds, 23

Black Hole Production by Cosmic Rays, 50

Black hole scan, 41

Black hole solutions of two-dimensional Riemann–Cartan gravity with higher-derivative action, 85

Black hole superpartners and fixed scalars, 68

Black hole thermodynamics and two-dimensional dilaton gravity theory, 111

Black Hole Thermodynamics from the Point of View of Superstring Theory, 14

Black holes and causal structure in anti-de Sitter isometric spacetimes, 62

Black Holes and Conformal Mechanics, 85

Black holes and energy pirates: how to recognize and release them, 7

Black holes and flop transitions in M-theory on Calabi-Yau 3-folds, 53

Black holes and instabilities of negative tension branes, 83

Black holes and naked singularities in the low-energy limit of string gravity with a modulus field, 15

Black holes and relativistic stars, 1

Black holes and supernovae, 6

Black holes and the SYM phase diagram, 77, 83

Black holes and the SYM phase diagram. II, 83

Black holes and wormholes in dimensions, 15

Black Holes at the Large Hadron Collider, 46

Black holes in nonflat backgrounds: The Schwarzschild black hole in the Einstein universe, 87

Black holes in supergravity and string theory, 86

Black holes in the brane world: Time symmetric initial data, 99

Black holes in three-dimensional Einstein-Born-Infeld-dilaton theory, 110

Black Holes of a Minimal Size in String Gravity, 14

Black holes of $D = 5$ supergravity, 54

Black holes of negative mass, 82

Black holes on thick branes, 48

Black Holes Radiate Mainly on the Brane, 48

Black holes versus naked singularities formation in collapsing Einstein clusters, 66

Black holes with less entropy than $A/4$, 89

Black holes with polyhedral multi-string configurations, 51

Black holes with topologically nontrivial AdS asymptotics, 16

Black holes with zero mass, 88

Black holes, 1, 2, 5, 6, 7, 8, 9, 15, 16, 48, 51, 53, 54, 62, 64, 66, 77, 82, 83, 86, 87, 88, 89, 99, 110

Black holes, gravitational radiation, and the universe: essays in honor of C.V. Vishveshwara, 1

Black Holes, Information Puzzle and String Theory, 107

Black holes, wormholes & time machines, 1

Black Holes: the end of the universe?, 9

Black strings and classical hair, 63

Blueshifted Quasars Associated with Nearby Galaxies?, 21

Bound states of branes with minimal energy, 67, 68

Boundary conditions and quasilocal energy in the canonical formulation of all 1+1 models of gravity, 74

Boundary terms and Noether current of spherical black holes, 16

Bounds from primordial black holes with a near critical collapse initial mass function, 73

Bounds on the cosmological abundance of primordial black holes from diffuse sky brightness: Single mass spectra, 42

Bounds on the inner radius of emission around supermassive black holes, 29

Brane-world black holes, 36

Brane-world creation and black holes, 54

Bulk and boundary dynamics in BTZ black holes, 70

C

Calabi-Yau black holes, 99

Calculation of the Emergent Spectrum and Observation of Primordial Black Holes, 61

Can Black Holes Decay to Naked Singularities?, 44

Can the effective string see higher partial waves?, 58

Capture of bulk geodesics by brane-world black holes, 36

Cardy-Verlinde formula and AdS black holes, 33

Casimir effect in 2D stringy black hole backgrounds, 38

Cauchy horizon stability in two-dimensional 'accelerated' black holes, 49

Cet objet obscur: le trou noir, 75

Chandra Uncovers a Hidden Low-Luminosity Active Galactic Nucleus in the Radio Galaxy Hydra A (3C 218), 96

Chaos in black holes surrounded by gravitational waves, 76

Charged and rotating AdS black holes and their CFT duals, 60

Charged Black Holes in Asymptotically Anti-deSitter Spacetime, 85

Charged brane-world black holes, 37

Charged rotating black hole in three spacetime dimensions, 83

Charged Sectors, Spin and Statistics in Quantum Field Theory on Curved Spacetimes, 58

Classical and quantum black holes, 2

Classical and Quantum Black Holes, 51

Classical super-radiance in Kerr-Newman-anti-de Sitter black holes, 109

Classicality of Primordial Fluctuations and Primordial Black Holes, 92

Co-accelerated particles in the C-metric, 92

Coalescence of two spinning black holes: An effective one-body approach, 43

Collapsing shells and the isoperimetric inequality for black holes, 55

Collision of spinning black holes in the close limit, 70

Comment on, 59, 79, 90, 100

Comments on black holes in matrix theory, 63

Comments on black holes in string theory, 63

Comparing D-branes and black holes with 0- and 6-brane charges, 91

Complete null data for a black hole collision, 56

COMPTEL all-sky imaging at 2.2 MeV, 84

Computing radiation from Kerr black holes: Generalization of the Sasaki-Nakamura equation, 64

Conformal anomaly for 2D and 4D dilaton coupled spinors, 105

Conformal dynamics of 0-branes, 31

Conformal Superfields and BPS States in AdS4/7 Geometries, 50

Constraints on the mass and abundance of black holes in the Galactic halo: the high-mass limit, 87

Contribution of High-Mass Black Holes to Mergers of Compact Binaries, 24

Correspondence principle for black holes and strings, 63

Cosmic rays from remnants of quasars?, 26

Cosmic String Loops Collapsing to Black Holes, 59

Cosmological evolution of black holes in Brans-Dicke gravity, 95

Cosmological supergravity from a massive superparticle and super-cosmological black holes, 72

Critical behavior and universality in gravitational collapse of a charged scalar field, 62

D

D3-brane shells to black branes on the Coulomb branch, 56
D-branes and near extremal black holes at low energies, 81
D-branes, moduli, and supersymmetry, 18
Density fluctuations and primordial black hole formation in natural double inflation in supergravity, 110
Detecting Energy Emissions from a Rotating Black Hole, 105
Detection of variable frequency signals using a fast chirp transform, 66
Detection Techniques of Microsecond Gamma-Ray Bursts Using Ground-based Telescopes, 73
Diffractive/refractive optics for high energy astronomy, 67
Dilaton black holes on thick branes, 95
Dilaton gravity (with a Gauss-Bonnet term) derived from five-dimensional Chern-Simons gravity, 20
Dilaton test of connection between AdS $3 \times S\ 3$ and 5D black holes, 75
Dilatonic brane-world black holes, gravity localization, and Newton's constant, 88
Dilatonic monopoles and, 29
Dimensional reduction of 4D heterotic string black holes, 32
Discovery of a massive equatorial torus in the η Carinae stellar system, 46
Discovery of recurring soft-to-hard state transitions in LMC X-3, 109
Discrete quantum gravity and causal sets, 94
Distorted charged dilaton black holes, 110
Diverse supernova sources for the r-process, 93
Do collapsed boson stars result in new types of black holes?, 91
Do Semiclassical Zero Temperature Black Holes Exist?, 15
Do stringy corrections stabilize colored black holes?, 69
Does brane cosmology have realistic principles?, 41
Duality invariance of black hole creation rates, 30
Dynamic and thermodynamic stability and negative modes in Schwarzschild-anti-de Sitter black holes, 93
Dynamical black holes in scalar-tensor theories, 70
Dynamics of black hole motion, 42
Dynamics of black holes in galactic centres, 4
Dynamics of Plasma Close to the Horizon of a Schwarzschild Black Hole, 38
Dyonic BIon black hole in string inspired model, 103
Dyonic non-Abelian black holes, 29

E

Editorial, 97
Effect of sources on the inner horizon of black holes, 58
Effect of spin on the quantum entropy of black holes, 67
Effect of violation of quantum mechanics on neutrino oscillation, 78
Effective action and Hawking radiation for dilaton coupled scalars in two dimensions, 74
Effective field theory of slowly moving `extreme black holes', 45
Effective Field Theory, Black Holes, and the Cosmological Constant, 39
Effective Spatial Dimension of Extremal Nondilatonic Black p-Branes and the Description on the World Volume, 32
Effective theories and black hole production in warped compactifications, 55
Effects of Kerr space-time on spectral features from X-ray illuminated accretion discs, 83
Effects of non-thermal tails in Maxwellian electron distributions on synchrotron and Compton processes, 108
Eikonal particle scattering and dilaton gravity, 44
Einstein's relativity & the quantum revolution [sound recording], 12
Einstein-Yang-Mills isolated horizons: Phase space, mechanics, hair, and conjectures, 40
Electric charge and magnetic flux on rotating black holes in a force-free magnetosphere, 76
Electrically Charged Einstein-Born-InfeldBlack Holes with Massive Dilaton, 110
Electromagnetic waves in a strong Schwarzschild plasma, 43
Electron-Positron Outflow from Black Holes, 105
Energy and angular momentum flow into a black hole in a binary, 15
Enhancement of supersymmetry near a 5D black hole horizon, 37
Entropies of rotating charged black holes from conformal field theory at Killing horizons, 67
Entropy bound for a charged rotating system, 93
Entropy bound for a rotating system from anti-de Sitter black holes, 107

Entropy bound of a charged object and electrostatic self-energy in black holes, 78
Entropy for asymptotically AdS 3 black holes, 87
Entropy from conformal field theory at Killing horizons, 34
Entropy of 4D black holes and the enhanc,on, 40
Entropy of four-dimensional rotating BPS black hole dyons, 54
Entropy of localized states and black hole evaporation, 89
Entropy of N=2 black holes and their M-brane description, 22
Entropy of near-extreme N=2 black holes, 22
Entropy of very low energy localized states, 89
Equation of state for a classical gas of BPS black holes, 69
Equilibrium and stability of supermassive stars in binary systems, 99
Erratum: Rotating topological black holes [Phys. Rev. D 57, 6127 (1998)], 71
Estimating Hawking radiation for exotic black holes, 102
Evaporation of Near-Extremal Reissner-Nordström Black Holes, 49
Evidence for a null entropy of extremal black holes, 62
Evidence for ionized accretion discs in five narrow-line Seyfert 1 galaxies, 19
Evidence for Supermassive Black Holes in Active Galactic Nuclei from Emission-Line Reverberation, 91
Evolution of Binary Compact Objects That Merge, 24
Evolution of circular, nonequatorial orbits of Kerr black holes due to gravitational-wave emission, 65
Evolution of massive binary black holes, 111
Evolution of Proto-Neutron Stars with Quarks, 92
Exact black hole entropy bound in conformal field theory, 26
Existence of stable hairy black holes in su(2) Einstein-Yang-Mills theory with a negative cosmological constant, 109
Exploring black holes: introduction to general relativity, 9
Extendability of Solutions of the Einstein-Yang/Mills Equations, 101
Extremal Black Holes and the Limits of the Third Law, 77
Extreme Black Hole Entropy Obtained in an Operational Approach, 108
Extreme Kerr throat geometry: A vacuum analog of AdS 2 ×S 2, 21

F

Facts on File stars & planets atlas, 7
Fate of a Reissner-Nordström black hole in the Einstein-Yang-Mills-Higgs system, 103
Fate of the Black String Instability, 64
Fixed scalar greybody factors in five and four dimensions, 71
Fixed scalars and suppression of Hawking evaporation, 72
Flash! : the hunt for the biggest explosions in the universe, 7
Formation and evaporation of charged black holes, 101
Formation Rates of Black Hole Accretion Disk Gamma-Ray Bursts, 52
Friedmann Equation and Cardy Formula Correspondence in BraneUniverses, 107
From Computation to Black Holes and Space-Time Foam, 87
From quarks to black holes: interviewing the universe, 3
From white dwarfs to black holes: the legacy of S. Chandrasekhar, 3

G

Gamma-Ray Bursts as the Birth-Cries of Black Holes, 67
Gauge-invariant perturbations of Schwarzschild black holes in horizon-penetrating coordinates, 96
General rotating black holes in string theory: Greybody factors and event horizons, 42
Geometry and topology of two kinds of extreme Reissner-Nordström-anti-de Sitter black holes, 108
Geons, black holes, and quantum foam: a life in physics, 11
Global embeddings of scalar-tensor theories in 2+1 dimensions, 62
Global structure of Robinson-Trautman radiative space-times with cosmological constant, 25
Gott Time Machines, BTZ Black Hole Formation, and Choptuik Scaling, 26
Gravitating BIon and BIon black hole with a dilaton, 103
Gravitating discs around a Schwarzschild black hole: I, 98
Gravitating monopoles and black holes at intermediate Higgs boson masses, 29
Gravitating sphalerons and sphaleron black holes in asymptotically anti-de Sitter spacetime, 105

Gravitational collapse in 2+1 dimensional AdS spacetime, 93
Gravitational collapse in a constant potential bath, 66
Gravitational collapse of gravitational waves in 3D numerical relativity, 14
Gravitational Collapse of Spherically Symmetric Perfect Fluidwith Kinematic Self-Similarity, 28
Gravitational lensing in eclipsing binary stars, 83
Gravitational properties of monopole spacetimes near the black hole threshold, 79
Gravitational scattering of cosmic strings by non-rotating black holes, 106
Gravitational Thermodynamics and Black-Hole Mergers, 92
Gravitational ultrarelativistic interaction of classical particles in the context of unification of interactions, 92
Gravitational wave astronomy, 97
Gravitational wave chirp search: Economization of post-Newtonian matched filter bank via cardinal interpolation, 41
Gravitational waves from galactic centres?, 93
Gravitational waves from quasispherical black holes, 60
Gravitational waves, black holes and cosmic strings in cylindrical symmetry, 60, 95
Gravitational-wave dynamics and black-hole dynamics: second quasi-spherical approximation, 60
Gravity, parametric resonance, and chaotic inflation, 47
Grazing Collisions of Black Holes via the Excision of Singularities, 28
Greybody factors for black holes in four dimensions: Particles with spin, 42
Growth of stellar mass black holes in galactic nuclei, 76

H

Hairy black holes, horizon mass and solitons, 16
Hawking and black holes, 8
Hawking radiation from AdS black holes, 61
Hawking radiation from four-dimensional Schwarzschild black holes in M theory, 43
Hawking radiation in string theory and the string phase of black holes, 93
Hawking Radiation without Black Hole Entropy, 107
Hermitian D-brane solutions, 85

High energy effects on D-brane and black hole emission rates, 44
High-energy QCD as a topological field theory, 48
Higher dimensional flat embeddings of (2+1)-dimensional black holes, 63
Higher dimensional flat embeddings of black strings in 2+1 dimensions, 63
Higher dimensional flat embeddings of black strings in, 62, 63
Higher dimensional inhomogeneous dust collapse and cosmic censorship, 55
Higher dimensional Kerr-AdS black holes and the AdS/CFT correspondence, 17
Holographic particle detection, 19
Holographic probes of anti–de Sitter spacetimes, 19
Holographic stress tensors for Kerr-AdS black holes, 17
Holography and brane cosmology in domain wall backgrounds, 33
Holography, thermodynamics, and fluctuations of charged AdS black holes, 37
Horizon dynamics of evaporating black holes in a higher dimensional inflationary universe, 84
Horizon holography, 95
Horizon/matter systems near the extreme state, 112
Hot Plasma and Black Hole Binaries in Starburst Galaxy M82, 57
How frustrated strings would pull the black holes from the centers of galaxies, 101
How to avoid artificial boundaries in the numerical calculation of black hole spacetimes, 64
Hunting down the universe: the missing mass, primordial black holes, and other dark matters, 3, 4

I

Impacts of the Detection of Cassiopeia A Point Source, 104
Imprints of accretion on gravitational waves from black holes, 90
In defense of the "tunneling" wave function of the universe, 54
Increase of the Hawking Radiation for Spinor Particles from Schwarzschild Black Holes by Dirac Monopoles, 56
Infinitely coloured black holes, 84
Inflationary preheating and primordial black holes, 21

Initial data and spherical dust collapse, 85
Initial data and the end state of spherically symmetric gravitational collapse, 67
Initial data for a head-on collision of two Kerr-like black holes with close limit, 43
Initial Data for Two Kerr-like Black Holes, 43
Innermost stable circular orbit of binary black holes, 22
Inspiraling Black Holes: The Close Limit, 70
Instability of extremal relativistic charged spheres, 15
Instability of two-dimensional heterotic stringy black holes, 18
Integrability of the minimal strain equations for the lapse and shift in 3+1 numerical relativity, 56
Integrability of the N-Body Problem in (2+1)-AdS Gravity, 104
Integrable models and degenerate horizons in two-dimensional gravity, 41
Integrable models of galactic discs with double nuclei, 66
Interaction between black holes and dark matter, 74
Interpolating black holes in dilaton - axion gravity, 53
Introduction to plasma astrophysics, 89
Investigation of Radiative Outflows around Compact Objects, 37
Iron K line profiles driven by non-axisymmetric illumination, 111
Iron Opacity and the Pulsar of SN 1987A, 52
Is Equilibrium of Aligned Kerr Black Holes Possible?, 82
Isolated horizons: Hamiltonian evolution and the first law, 16
Isometric embeddings of the reduced horizon of a Kerr-Newman black hole into Euclidean 3-space, 102

K

Kerr-AdS and Kerr-dS solutions reexamined, 14
Kinematics from spectroscopy with a wide slit: detecting black holes in galaxy centres, 80
Kruskal coordinates as canonical variables for Schwarzschild black holes, 105

L

Late time decay of scalar, electromagnetic, and gravitational perturbations outside rotating black holes, 20
Late-time decay of gravitational and electromagnetic perturbations along the event horizon, 20
Late-Time Decay of Scalar Perturbations Outside Rotating Black Holes, 21
Laws governing isolated horizons: inclusion of dilaton couplings, 16
Light rays at optical black holes in moving media, 29
Limits on black hole formation from cosmic string loops, 79
Line Emission from an Accretion Disk Around a Black Hole: Effects of Disk Structure, 90
LISA - an ESA cornerstone mission for a gravitational wave observatory, 103
Looking for event horizons using UV-IR relations, 57
Lorentzian approach to black hole thermodynamics in the Hamiltonian formulation, 27
Low-energy sector quantization of a massless scalar field outside a Reissner-Nordström black hole and static sources, 35
Low-frequency gravitational waves from cosmological compact binaries, 97
Low-frequency scalar absorption cross sections for stationary black holes, 61
Luminous hot accretion discs, 111

M

Magnetohydrodynamic interpretations of high-energy phenomena in galaxies and clusters, 81
Magnetohydrodynamical non-radiative accretion flows in two dimensions, 102
Mass formula for Einstein-Yang-Mills solitons, 41
Mass function of dormant black holes and the evolution of active galactic nuclei, 96
Mass of colored black holes, 41
Mass outflow rate from accretion discs around compact objects, 44
Massless scalar fields and topological black holes, 46
Matching Condition on the Event Horizon and Holography Principle, 47
Matrix black holes, 77
Mechanics of rotating isolated horizons, 16
Mechanism of the generation of black hole entropy in Sakharov's induced gravity, 52
Membranes wrapped on holomorphic curves, 54
Microcanonical statistics of black holes and the bootstrap condition, 64

Microfield dynamics of black holes, 35
Microscopic black-hole pairs in highly excited states, 80
Microscopic entropy of N=2 extremal black holes, 69
Million solar mass black holes at high redshift, 56
Mining energy from a black hole by strings, 51
Model dependence of transonic properties of accretion flows around black holes, 36
Moduli spaces for four- and five-dimensional black holes, 58
Monopoles, dyons, and black holes in the four-dimensional Einstein-Yang-Mills theory, 26
More D-brane bound states, 29
Mother templates for gravitational wave chirps, 96
Multicritical phenomena of Reissner-Nordström anti-de Sitter black holes, 110
Mysteries of the universe, 11

N

Naked black holes, 63
Naked Reissner-Nordström singularities and the anomalous magnetic moment of the electron field, 23
Naked singularities in higher dimensional Vaidya space-times, 55
Naked singularities in low-energy, effective string theory, 70
Nearly horizon skimming orbits of Kerr black holes, 65
Neutrino fluxes from active galaxies: A model-independent estimate, 59
Neutron Stars and Black Holes as MACHO, 106
New angles on D-branes, 29
New critical behavior in Einstein-Yang-Mills collapse, 38
New five-dimensional black holes classified by horizon geometry, and a Bianchi VI braneworld, 31
New Hypointense Lesions on MRI in Relapsing-Remitting Multiple Sclerosis Patients, 107
Newtonian versus black-hole scattering, 100
Noether currents of charged spherical black holes, 16
No-go theorem for false vacuum black holes, 53
Noise kernel in stochastic gravity and stress energy bitensor of quantum fields in curved spacetimes, 91
Noise-driven evolution in stellar systems - II. A universal halo profile, 108
Non-Abelian Einstein-Born-Infeld black holes, 109
Nonextremal stringy black hole, 102
Non-extreme black holes near the extreme state and acceleration horizons: thermodynamics and quantum-corrected geometry, 112
Non-Gaussian fluctuations and primordial black holes from inflation, 30
Nonlinear and perturbative evolution of distorted black holes: Odd-parity modes, 18
Nonlinear evolution of de Sitter space instabilities, 87
Non-linear evolution of thermally unstable slim accretion discs with a diffusive form of viscosity, 102
Nonthermal nature of incipient extremal black holes, 77
Nonthreshold D-brane bound states and black holes with nonzero entropy, 41
Nonvanishing magnetic flux through the slightly charged Kerr black hole, 71
Note on the entropy of charged multi-black-holes, 86
Numerical calculation of conformally smooth hyperboloidal data, 64
Numerical evidence against black holes with non-minimally coupled scalar hair, 91
Numerical testbed for singularity excision in moving black hole spacetimes, 111

O

Object picture of quasinormal ringing on the background of small Schwarzschild anti–de Sitter black holes, 113
Observing the Birth of Supermassive Black Holes with the Planned ICECUBE Neutrino Detector, 98
On domain-wall/QFT dualities in various dimensions, 22
On some transonic aspects of general relativistic spherical accretion on to Schwarzschild black holes, 44
On the complex disc-corona interactions in the soft spectral states of soft X-ray transients, 112
On the detectability of distant Compton-thick obscured quasars, 49
On the entropy of matrix black holes, 77
On the job with an astronomer: explorer of the universe, 5
On the linear stability of solitons and hairy black holes with a negative cosmological constant: the odd-parity sector, 96

On the Oppenheimer-Volkoff Equations in General Relativity, 101

On the peak radio and X-ray emission from neutron star and black hole candidate X-ray transients, 49

On the Role of Minor Galaxy Mergers in the Formation of Active Galactic Nuclei, 40

On the topology and area of higher-dimensional black holes, 32

One-loop corrected thermodynamics of the extremal and nonextremal spinning Banados-Teitelboim-Zanelli black hole, 84

Open strings, 2D gravity, and AdS/CFT correspondence, 32

Optimal entropy bound and the self-energy of test objects in the vicinity of a black hole, 84

Orbital Comptonization in accretion disks around black holes, 94

Oscillatory Null Singularity inside Realistic Spinning Black Holes, 89

Outbursts from a Black Hole via Alfvén Wave to Electromagnetic Wave Mode Conversion, 43

P

Pacific Conference on Gravitation and Cosmology: February 1-6, 1996, Sheraton Walker-Hill, Seoul, Korea, 6

Pair creation of dilaton black holes in extended inflation, 28

Pair production of topological anti-de Sitter black holes, 82

Particle Acceleration and Curvature TeV Emission by Rotating, Supermassive Black Holes, 76

Particle definition in the presence of black holes, 98

Perturbation theory for self-gravitating gauge fields: The odd-parity sector, 96

Perturbations in the Kerr-Newman dilatonic black hole background: Maxwell waves, 35

Perturbative evolution of nonlinear initial data for binary black holes: Zerilli versus Teukolsky equation, 78

Perturbing supersymmetric black holes, 89

Phase Lag and Coherence Function of X-Ray Emission from Black Hole Candidate XTE J1550-564, 41

Phase Transitions for Flat Anti–de Sitter Black Holes, 102

Plunge Waveforms from Inspiralling Binary Black Holes, 18

Positive mass theorem for black holes in Einstein-Maxwell axion-dilaton gravity, 95

Pragmatic Approach to Gravitational Radiation Reaction in Binary Black Holes, 78

Precursors, black holes, and a locality bound, 55

Primordial Black Hole Evolution in Tensor-Scalar Cosmology, 65

Primordial black hole formation during the QCD epoch, 66

Primordial black hole production due to preheating, 57

Primordial black holes and primordial nucleosynthesis: Effects of hadron injection from low mass holes, 72

Primordial black holes from inflationary models with and without broken scale invariance, 29

Primordial black holes under the double inflationary power spectrum, 71

Probability for primordial black holes in a higher dimensional universe, 90

Probing black holes in nonperturbative gauge theory, 65

Probing Microquasars with TeV Neutrinos, 77

Probing near extremal black holes with D-branes, 81

Production of intermediate-mass black holes in globular clusters, 39

Proof of the generalized second law for two-dimensional black holes, 62

Properties of black hole solutions in the SU(3) Einstein-Yang-Mills-dilaton system, 103

Proton Decay, Black Holes, and Large Extra Dimensions, 14

Q

QED blue-sheet effects inside black holes, 31

Quanta of geometry and rotating black holes, 72

Quantization of maximally charged slowly moving black holes, 100

Quantization of the electromagnetic field outside static black holes and its application to low-energy phenomena, 41

Quantum back reaction of massive fields and self-consistent semiclassical extreme black holes and acceleration horizons, 84

Quantum black holes are tied to D-branes and strings, 39, 40

Quantum black holes from quantum collapse, 105

Quantum corrections to the entropy of a Reissner-Nordström black hole due to spin fields, 77

Quantum Corrections to the Entropy of ExtremeReissner-Nordström Black Holes, 93
Quantum corrections to the thermodynamics of charged 2D black holes, 84
Quantum entropy of a nonextreme stationary axisymmetric black hole due to a minimally coupled quantum scalar field, 67
Quantum geometry and thermal radiation from black holes, 73
Quantum gravity corrections for Schwarzschild black holes, 22
Quantum gravity: proceedings of the sixth Moscow seminar: Moscow, Russia, June 12-19, 1995, 6
Quantum instability of two-dimensional charged black holes, 58
Quantum mechanical path integrals and thermal radiation in static curved spacetimes, 106
Quantum minimal length and trans-Planckian photons, 79
Quantum probes of repulsive singularities in supergravity, 53
Quantum radiation from black holes and naked singularities in spherical dust collapse, 99
Quantum states and the statistical entropy of the charged black hole, 105, 106
Quantum, Gravity and Geometry, 99
Quantum-mechanical model of the Kerr-Newman black hole, 81
Quasars and galaxy formation, 36
Quasicircular orbits for spinning binary black holes, 91
Quasilocal energy for rotating charged black hole solutions in general relativity and string theory, 28
Quasilocal thermodynamics of Kerr and Kerr--anti-de Sitter spacetimes and the AdS/CFT correspondence, 45
Quasinormal modes of AdS black holes and the approach to thermal equilibrium, 64
Quasi-normal modes of AdS black holes: a superpotential approach, 57
Quasinormal modes of charged, dilaton black holes, 50
Quasinormal modes of Schwarzschild-anti-de Sitter black holes: Electromagnetic and gravitational perturbations, 34
Quasinormal modes: the characteristic 'sound' of black holes and neutron stars, 88
Quasispherical approximation for rotating black holes, 99
Question of Abelian-Higgs hair expulsion from extremal dilaton black holes, 86
Quintessence with a Localized Scalar Field on the Brane, 87

R

Radiation from a class of string theoretic black holes, 45, 95
Radiation recoil from highly distorted black holes, 28
Radiative falloff in neutron star spacetimes, 90
Rates of tidal disruption of stars by massive central black holes, 80
Redshifts near black holes, 61
Regular (2+1)-dimensional black holes within nonlinear electrodynamics, 35
Regular black holes and topology change, 27
Regular magnetic black holes and monopoles from nonlinear electrodynamics, 30
Regular R-R and NS-NS BPS Black Holes, 24
Reissner–Nordström-like solutions of the SU(2) Einstein–Yang/Mills equations, 101
Relativistic Viscous Fluid Description of Microscopic Black Hole Wind, 69
Remark on the formation of colored black holes via fine-tuning, 26
Remarks on Dirichlet branes at angles, 59
Renormalized black hole entropy in anti-de Sitter space via the, 109
Renormalized thermodynamic entropy of black holes in higher dimensions, 71
Reply to, 32
Representations of Superconformal Algebras in the AdS 7/4/CFT 6/3 Correspondence, 50
Resolution of cosmological singularities in string theory, 74, 75
Resolving the Black Hole Information Paradox, 83
Reviews in modern astronomy 14
Reviews in modern astronomy: dynamic stability and instabilities in the universe, 7
Rossi X-Ray Timing Explorer Observations of an Outburst of Recurrent X-Ray Nova GS 1354-644, 94
Rotating Hairy Black Holes, 71

S

S -Matrices for Quantum Charged Massive Scalar Particles on Schwarzschild and Reissner-Nordström Black Holes, 51
Scalar absorption by spinning D3-branes, 100
Scalar, electromagnetic, and Weyl perturbations of BTZ black holes: Quasinormal modes, 34

Scale versus conformal invariance in the AdS/CFT correspondence, 17
Scales of gravity, 47
Scattering in three-dimensional extremal black holes, 54
Scattering map for two black holes, 45
Scattering of Dirac waves off Kerr black holes, 36
Scattering of four-dimensional black holes, 85
Scattering of scalar waves by rotating black holes, 56
Schwarzschild black hole as a grand canonical ensemble, 57
Schwarzschild Black Holes from Matrix Theory, 20
Selection Rules for Splitting Strings, 13
Self-consistent axisymmetric Sridhar-Touma models, 66
Self-force on a static scalar test charge outside a Schwarzschild black hole, 109
Semiclassical charged black holes with a quantized massive scalar field, 103
Singularity dynamics in curvature collapse and jet eruption on a fluid surface, 113
Small-scale inviscid accretion discs around black holes, 23
Soft state of Cygnus X-1: stable disc and unstable corona, 38
Solitons and black holes in Einstein-Born-Infeld-dilaton theory, 39
Solving the Initial Value Problem of Two Black Holes, 83
Some comments about Schwarzschild black holes in matrix theory, 72
Some comments on gravitational entropy and the inverse mean curvature flow, 55
Sonic Analog of Gravitational Black Holes in Bose-Einstein Condensates, 54
Sonic black holes in dilute Bose-Einstein condensates, 54
Space & art activities, 3
Space mysteries, 6
Space, 2, 3, 6
Spectral variability in transonic discs around black holes, 112
Spectrum of charged black holes—the big fix mechanism revisited, 21
Spherically symmetric false vacuum: No-go theorems and global structure, 30
Spinning string fluid dynamics in general relativity, 100
Stability and collapse of rapidly rotating, supramassive neutron stars: 3D simulations in general relativity, 98
Stability of AdS and phase transitions, 60
Stability of inflationary solutions driven by a changing dissipative fluid, 38
Stable black strings in anti-de Sitter space, 61
Stars and black holes in varying speed of light theories, 80
Stars, 11
Static and infalling quasilocal energy of charged and naked black holes, 27
Static and rotating electrically charged black holes in three-dimensional Brans-Dicke gravity theories, 46
Static black holes of metric-affine gravity in the presence of matter, 17
Statistical Entropy of Four-Dimensional Rotating Black Holes from Near-Horizon Geometry, 42
Statistical entropy of Schwarzschild black strings and black holes, 44
Statistical Origin of Black Hole Entropy in Matrix Theory, 79
Stellar remnants, 4
Strange stuff: true stories of odd places and things, 5
Strength of the null singularity inside black holes, 31
String excitation inside generic black holes, 80
String model of black hole microstates, 74
String propagation in an exact four-dimensional black hole background, 80
String versus Einstein frame in an AdS/CFT induced quantum dilatonic brane-world universe, 88
Strings, branes, and gravity: TASI 99: Boulder, Colorado, USA, 31 May-25 June 1999, 8
Stringy black holes and energy conditions, 69
Structure of the Black Hole[close quote]s Cauchy-Horizon Singularity, 31
Structure of the Black Hole's Cauchy-Horizon Singularity, 31
Submillimeter Evidence for the Coeval Growth of Massive Black Holes and Galaxy Bulges, 89
Summation over histories for a particle in spherical orbit around a black hole, 23, 24
Supergravity description of non-Bogomol'nyi-Prasad-Sommerfield branes, 29
Superradiance Resonance Cavity Outside Rapidly Rotating Black Holes, 15
Superstrings, Gauge Fields and Black Holes, 18

Supersymmetric conical defects: Towards a string theoretic description of black hole formation, 19

Supersymmetric rotating black holes and attractors, 68

Supersymmetric rotating black holes and causality violation, 52

Supersymmetric solutions to topologically massive gravity and black holes in three dimensions, 45

Surface counterterms and boundary stress-energy tensors for asymptotically non-anti-de Sitter spaces, 33

Surface terms as counterterms in the AdS-CFT correspondence, 48

T

Testing effective string models of black holes with fixed scalars, 72

The 0-brane action in a general $D = 4$ supergravity background, 25

The amazing space almanac, 7

The area spectrum in quantum gravity, 73

The Bekenstein formula and string theory (N-brane theory), 90

The black hole: 25 years after, 9

The Cardy-Verlinde Formula and Asymptotically Flat Charged Black Holes, 111

The classical stability of charged $(1 + 1)$-dimensional black holes, 75

The correlation between black hole mass and bulge velocity dispersion in hierarchical galaxy formation models, 59

The effects of a Comptonizing corona on the appearance of the reflection components in accreting black hole spectra, 91

The enhanc,on, black holes, and the second law, 67

The first law of black-hole thermodynamics for black holes in string theory, 98

The formation of supermassive black holes and the evolution of supermassive stars, 87

The Galactic Center: An Interacting System of Unusual Sources, 112

The gravitating model in $2 + 1$ dimensions: black hole solutions, 39

The holographic principle for general backgrounds, 28

The influence of central black holes on gravitational lenses, 82

The interplay between forces in the Kerr - Newman field, 98

The knowledge network, 9

The mystery of black holes, 6

The neutron star-black hole connection, 9

The obscured growth of massive black holes, 49

The physical properties of charged five-dimensional black holes, 78

The Radiative Feedback of the First Cosmological Objects, 59

The Reader's Digest children's atlas of the universe, 9

The relationship between X-ray variability and the central black hole mass, 79

The SAURON project - II. Sample and early results, 104

The Scientific American book of the cosmos, 10

The sixth Canadian Conference on General Relativity and Relativistic Astrophysics, 10

The staticity problem for non-rotating black holes in Einstein - Maxwell axion - dilaton gravity, 95

The theory of everything: the origin and fate of the universe, 3

The two-phase approximation for black hole collisions: is it robust?, 18

The universe., 11

The World in Eleven Dimensions, 47

Theory of black hole accretion disks, 11

Thermal equilibria of accretion discs around black holes: a review of the recent progress, 13

Thermal fields, entropy and black holes, 52

Thermodynamic properties of massive dilaton black holes, 103

Thermodynamics of Kerr-Newman-AdS black holes and conformal field theories, 33

Thermodynamics of Schwarzschild-(Anti-)de Sitter Black Holes with Account of Quantum Corrections, 88

Three-dimensional black hole from a stringy anti–de Sitter background, 62

Tidal Interaction in Binary-Black-Hole Inspiral, 93

Topological black holes in anti-de Sitter space, 25

Topological black holes in Weyl conformal gravity, 71

Topological censorship and higher genus black holes, 53

Toward a midisuperspace quantization of LeMaitre-Tolman-Bondi collapse models, 106

Towards a stable numerical evolution of strongly gravitating systems in general relativity: The conformal treatments, 14

Towards a Theory of Quantum Black Holes, 23

Towards the solution of the relativistic gravitational radiation reaction problem for binary black holes, 78

Transition from inspiral to plunge in binary black hole coalescences, 30

Truncation of geometrically thin disks around massive blackholes in galactic nuclei, 48

Twisted sectors in three-dimensional gravity, 20

Two black hole holography, lensing, and intensity, 65

Two Kerr black holes with axisymmetric spins: head-on collision and gravitational radiation, 15

Two-dimensional black holes in accelerated frames: quantum aspects, 19

U

Undergraduate thermodynamics and black holes, 47

Understanding the chiral anomaly in coordinate space, 101

Unified first law of black-hole dynamics and relativistic thermodynamics, 61

Uniformly accelerated black holes, 76

Uniformly accelerating black holes in a de Sitter universe, 92

Uniqueness theorems for static black holes in metric-affine gravity, 17

Universal low-energy dynamics for rotating black holes, 81

Unruh effect with back reaction: A first-quantized treatment, 94

Untangling the merger history of massive black holes with LISA, 65

V

Variations on conservation laws for the wave equation, 86

Viability of primordial black holes as short period gamma-ray bursts, 57

Vortices and black holes in dilatonic gravity, 96

W

Weighing black holes with warm absorbers, 86

What can we learn from the surface chemical composition of the optical companions of Soft X-ray transients?, 48

Why aren't black holes black? : the unanswered questions at the frontiers of science, 4

X

X-ray absorption and rapid variability of the dwarf Seyfert nucleus of NGC 4395, 65

X-ray Nova Binary Systems, 38

X-ray spectra of accretion discs with dynamic coronae, 81

GUIDE TO BOOKS

A science odyssey--short trip. Blips, black holes, and the Nobel Prize. Published/Created: United States: PBS-TV, 1998. Description: 1 videocassette of 1: sd., col. and b&w; 3/4 in. viewing copy. Notes: Copyright: WGBH Educational Foundation. DCR 1998; PUB 2Mar98; REG 3Jun98; PA891-545. Sources used: copyright data sheet; copyright data base. LC Classification: VBO 7826 (viewing copy)

Al-Khalili, Jim, 1962- Black holes, wormholes & time machines / Jim Al-Khalili. Published/Created: Bristol, UK; Philadelphia, PA: Institute of Physics Pub., c1999. Description: xxii, 265 p.: ill.; 22 cm. ISBN: 0750305606 (pbk.: alk. paper) Notes: Includes bibliographical references (p. 254-258) and index. Subjects: Space and time. LC Classification: QC173.59.S65 A4 1999 Dewey Class No.: 530.11 21

Black holes and relativistic stars / edited by Robert M. Wald. Published/Created: Chicago: University of Chicago Press, 1998. Related Names: Wald, Robert M. Description: xii, 278 p.: ill.; 24 cm. ISBN: 0226870340 (cloth: alk. paper) Notes: Includes bibliographical references. Subjects: Chandrasekhar, S. (Subrahmanyan), 1910- Black holes (Astronomy) LC Classification: QB843.B55 B585 1998 Dewey Class No.: 523.8/875 21

Black holes in binaries and galactic nuclei: diagnostics, demography, and formation: proceedings of the ESO workshop held at Garching, Germany, 6-8 September 1999, in honour of Riccardo Giacconi / L. Kaper, E.P.J. van den Heuvel, P.A. Woudt (eds.). Published/Created: Berlin; New York: Springer, c2001. Related Names: Giacconi, Riccardo. Kaper, Lex, 1966- Heuvel, Edward Peter Jacobus van den, 1940- Woudt, P. A. (Patrick A.), 1970- European Southern Observatory Description: xxiii, 378 p.: ill.; 25 cm. ISBN: 3540415815 (hardcover: acid-free paper) Notes: Includes bibliographical references and index. Subjects: Black holes (Astronomy)--Congresses. Double stars--Congresses. Galactic nuclei--Congresses. Series: ESO astrophysics symposia LC Classification: QB843.B55 B5873 2001 Dewey Class No.: 523.8/875 21

Black holes, gravitational radiation, and the universe: essays in honor of C.V. Vishveshwara / edited by Bala R. Iyer, Biplab Bhawal. Published/Created: Dordrecht; Boston: Kluwer, c1999. Related Names: Vishvshwara, C. V. Iyer, B. R. Bhawal, Biplab. Description: x, 565 p.: ill.; 25 cm. ISBN: 0792353080 (alk. paper) Notes: Includes bibliographical references. Subjects: Black holes (Astronomy) Gravitational waves. Series: Fundamental theories of physics; v. 100 LC Classification: QB843.B55 B587 1999 Dewey Class No.: 523.8/875 21

Black holes: theory and observation: proceedings of the 179th W.E. Heraeus Seminar, held at Bad Honnef, Germany, 18-22 August 1997 / Friedrich W. Hehl, Claus Kiefer, Ralph J.K. Metzler (eds.). Published/Created: Berlin; New York: Springer, c1998. Related Names: Hehl, F. W. (Friedrich W.), 1937- Kiefer, Claus, 1958- Metzler, Ralph J. K., 1969- Description: xv, 519 p.: ill.; 24 cm. ISBN: 3540651586 (acid-free paper) Notes: Includes bibliographical references and index. Subjects: Black holes (Astronomy)--Congresses. Astrophysics--Congresses. Series: Lecture notes in physics, 0075-8450; 514 LC Classification: QB843.B55 W4 1997 Dewey Class No.: 523.8/875 21

Classical and quantum black holes / Pietro Fré ... [et al.] (editors). Published/Created: Philadelphia: Institute of Physics Pub., 1999. Related Names: Fré, P. Description: xi, 348 p.: ill.; 25 cm. ISBN: 0750306270 (hbk.) Notes: Includes bibliographical references and index. Subjects: Black holes (Astronomy) Series: Studies in high energy physics, cosmology, and gravitation LC Classification: QB843.B55 C53 1999 Dewey Class No.: 523.8/875 21

Davis, Amanda. Black holes / Amanda Davis. Edition Information: 1st ed. Published/Created: New York: PowerKids Press, 1997. Description: 24 p.: ill. (some col.); 21 x 26 cm. ISBN: 0823950611 Summary: Briefly describes the formation and composition of black holes and the forces connected with them. Notes: Includes index. Subjects: Black holes (Astronomy)--Juvenile literature. Black holes (Astronomy) Series: Davis, Amanda. Exploring space. LC Classification: QB843.B55 D39 1997 Dewey Class No.: 523.8/875 21

Duality - strings & fields: proceedings of the 33rd Karpacz Winter School of Theoretical Physics, Karpacz, Poland, 13-22 February, 1997 / edited by Z. Hasiewicz, Z. Jaskólski, J. Sobczyk. Published/Created: [Amsterdam]: North-Holland, c1998. Related Names: Hasiewicz, Z. (Zbigniew) Jaskólski, Z. (Zbigniew) Sobczyk, Jan, 1955- Description: xii, 187 p.: ill.; 27 cm. Contents: Duality in string theory / S. Förste and J. Louis - Duality and global symmetries / F. Quevedo -- Aspects of N=1 string dynamics / S. Kachru -- Lectures on heterotic-type I duality / I. Antoniadis, H. Partouche and T.R. Taylor -- Properties of p-branes, D-branes and M-branes / E.A. Bergshoeff - Introduction to D-branes / L. Thorlacius -- Solitins, black holes and duality in string theory / R.R. Khuri -- Black holes and D-branes / J.M. Maldacena -- Classical and quantum composite p-branes / I. Ya. Aref'eva -- Massive and massless supersymmetric black holes / T. Ortin -- Duality and supersymmetric monopoles / J.P. Gauntlett -- Dualities in supersymmetric field theories / P.C. Argyres -- Novel field theory phenomena from F theory and D-branes / J. Sonnenschein -- M-theory, torons and confinement / C. Gómez and R. Hernández - WDVV equations in Seiberg-Witten theory and associative algebras / A. Mironov. Notes: Includes bibliographical references and author index. Subjects: Duality (Nuclear physics)--String models--Congresses. Field theory (Physics). Series: Nuclear physics. B, Proceedings, supplements; vol. 61A. LC Classification: QC770 .N772 vol. 61A Dewey Class No.: 539.7 s 530.14 21

Dyer, A. (Alan) Space / [author, Alan Dyer]. Published/Created: Pleasantville, N.Y.: Reader's Digest Children's Books, c1999. Description: 64 p.: col. ill.; 35 cm. ISBN: 1575842912 (hard cover) 1575842998 (lib. ed.) Summary: Examines the nature of outer space, our solar system, and our universe, including descriptions of the sun, moon, and planets, as well as constellations, nebulas and black holes. Notes: "A Weldon Owen production"--P. [2]. Includes index. Subjects: Astronomy--Outer space--Solar system--Juvenile literature. Series: Reader's Digest pathfinders LC Classification: QB500.22 .D94 1999 Dewey Class No.: 520 21

Frolov, V. P. (Valerii Pavlovich) Black hole physics: basic concepts and new developments / by Valeri P. Frolov and Igor D. Novikov. Published/Created: Dordrecht; Boston: Kluwer, c1998. Related Names: Novikov, I. D. (Igor' Dmitrievich) Description: xxi, 770 p.: ill.; 25 cm. ISBN: 0792351452 (acid free paper) Notes: Includes bibliographical references (p. 716-761) and index. Subjects: Black holes (Astronomy) Astrophysics. Series: Fundamental theories of physics; v. 96 LC Classification: QB843.B55 F76 1998 Dewey Class No.: 523.8/875 21

From white dwarfs to black holes: the legacy of S. Chandrasekhar / edited by G. Srinivasan. Published/Created: Chicago: University of Chicago Press, 1999. Related Names: Chandrasekhar, S. (Subrahmanyan), 1910- Srinivasan, G. (Ganesan), 1942- Description: xiii, 240 p.: ill.; 24 cm. ISBN: 0226769968 (alk. paper) Notes: "Essays ... originally published by the Indian Academy of Sciences in 1996 in a special issue of the Journal of astrophysics and astronomy"-- T.p. verso. Includes bibliographical references. Subjects: Chandrasekhar, S. (Subrahmanyan), 1910- Astronomy. Astrophysicists--United States-- Biography. LC Classification: QB51 .F734 1999 Dewey Class No.: 523.01/092 21

Goodman, Polly. Space & art activities / written by Polly Goodman. Published/Created: New York, NY: Crabtree Pub. Co., c2002. Description: p. cm. ISBN: 0778711404 (pbk.: alk. paper) 0778711129 (reinforced library binding: alk. paper) Summary: Information about various topics related to astronomy forms the foundation for projects about the sun and moon, planet rotation, space travel, black holes, and more. Contents: Solar system -- Blue world -- Spectacular sun -- Mysterious moon -- Mighty Jupiter -- Starry patterns -- The red planet -- Dust clouds -- Floating in space -- Saturn_s rings -- A star is born -- Galactic swirls -- Terrific telescopes -- Poison planet -- Rocks in space -- Spinning top -- Space walking -- Distant planets -- Trails of light -- Black holes -- Aliens. Subjects: Astronomy--Juvenile literature. Astronomy. Handicraft. Series: Arty facts LC Classification: QB46 .G69 2002 Dewey Class No.: 520 21

Hammond, Richard T. From quarks to black holes: interviewing the universe / Richard T. Hammond. Published/Created: River Edge, NJ: World Scientific, c2001. Description: xiii, 174 p.; 26 cm. ISBN: 9810246250 Subjects: Physics--Popular works. LC Classification: QC24.5 .H36 2001 Dewey Class No.: 530 21

Hawking, S. W. (Stephen W.) The theory of everything: the origin and fate of the universe / Stephen W. Hawking. Published/Created: Beverly Hills, CA: New Millennium Press, c2002. Description: p. cm. ISBN: 1893224546 (Hardcover) Contents: First lecture-ideas about the universe -- Second lecture-the expanding universe -- Third lecture-black holes -- Fourth lecture-black holes ain't so black -- Fifth lecture-the origin and fate of the universe -- Sixth lecture-the direction of time -- Seventh lecture-the theory of everything. Notes: Originally published: The Cambridge lectures: life works. West Hollywood, CA: Dove Books, c1996. Includes index. Subjects: Cosmology. Science--Philosophy. LC Classification: QB985 .H39 2002 Dewey Class No.: 523.1 21

Hawkins, Michael, 1942- Hunting down the universe: the missing mass, primordial black holes, and other dark matters / Michael Hawkins. Published/Created: Reading, Mass.: Addison-Wesley, c1997. Description: x, 240 p.; 24 cm. ISBN: 0201156989 (alk. paper) Notes: Includes bibliographical references (p. 219-222) and index. Subjects: Cosmology. Black holes (Astronomy) Dark matter (Astronomy) Astronomy--Philosophy. LC Classification: QB981 .H3775 1997 Dewey Class No.: 523.1 21

Hawkins, Michael. Hunting down the universe: the missing mass, primordial black holes, and other dark matters / Michael Hawkins, Celia Fitzgerald Hawkins. Published/Created: Reading, MA: Perseus Books, 1998. Description: p.; cm. ISBN: 0738200379 LC Classification: 9812 BOOK NOT YET IN LC

Hazen, Robert M., 1948- Why aren't black holes black?: the unanswered questions at the frontiers of science / Robert M. Hazen with Maxine Singer; foreword by Stephen Jay Gould. Edition Information: 1st Anchor Books trade pbk. ed. Published/Created: New York: Anchor Books, 1997. Related Names: Singer, Maxine. Description: xix, 309 p.; 21 cm. ISBN: 0385480148 (pbk.: alk. paper) Notes: Includes bibliographical references (p. [299]-309). Subjects: Science--Miscellanea. LC Classification: Q173 .H42 1997 Dewey Class No.: 500 21

Hemsendorf, Marc. Dynamics of black holes in galactic centres / Marc Hemsendorf. Published/Created: Aachen: Shaker, 2000. Description: iv, 109 p.: ill.; 21 cm. ISBN: 3826573552 (pbk.) Notes: Thesis (doctoral)--Rupertus Carola University, 1999. Includes bibliographical references (p. 89-96) Abstract also in German. Subjects: Black holes (Astronomy)--Mathematical models. Galactic nuclei--Mathematical models. Stellar dynamics. Series: Berichte aus der Astronomie, 0947-7756 LC Classification: QB843.B55 H45 2000 Dewey Class No.: 523.8/875/015118 21

High energy processes in accreting black holes: proceedings of a workshop held at Gräftåvallen, Sweden, 29 June-4 July, 1998 / edited by Juri Poutanen and Roland Svensson. Published/Created: San Francisco, Calif.: Astronomical Society of the Pacific, 1999. Related Names: Poutanen, Juri. Svensson, R. (Roland) Description: xxii, 449 p.: ill.; 24 cm. ISBN: 1886733813 Notes: Includes bibliographical references and index. Subjects: Black holes (Astronomy)--Accretion (Astrophysics)--X-ray astronomy--Gamma ray astronomy--Congresses. Series: Astronomical Society of the Pacific conference series; v. 161 LC Classification: QB843.B55 H54 1999 Dewey Class No.: 523.8/875 21

Internal structure of black holes and spacetime singularities: an international research workshop, Haifa, June 29-July 3, 1997 / edited on behalf of the Israel Physical Society by Lior M. Burko and Amos Ori. Published/Created: Bristol, UK; Philadelphia: Institute of Physics Pub.; Jerusalem: Israel Physical Society, c1997. Related Names: Burko, Lior M. Ori, Amos. Agudah ha-fisikalit le-Yísráel. Description: ix, 534 p.: ill.; 25 cm. ISBN: 0750305487 Notes: Includes bibliographical references. Subjects: Black holes (Astronomy)--Singularities (Mathematics)--Congresses. Space and time--Congresses. Series: Annals of the Israel Physical Society, 0309-8710; v. 13 LC Classification: QB843.B55 I57 1997 Dewey Class No.: 523.8/875 21

Kawaler, Steven D. Stellar remnants / S.D. Kawaler, I. Novikov, G. Srinivasan; edited by Georges Meynet and Daniel Schaerer. Published/Created: Berlin; New York: Springer, c1997 Related Names: Novikov, I. D. (Igor′ Dmitrievich) Srinivasan, G. (Ganesan), 1942- Meynet, G. Schaerer, Daniel. Description: x, 340 p.: ill.; 25 cm. ISBN: 3540615202 (acid-free paper) Notes: Includes bibliographical references and index. Subjects: Neutron stars--Congresses. White dwarfs--Congresses. Black holes (Astronomy)--Congresses. Stars--Evolution--Congresses. Series: Saas-Fee advanced course ... lecture notes; 1995. LC Classification: QB843.N4 K39 1997 Dewey Class No.: 523.8/87 20

Ketov, S. V. (Sergei Vladimirovich) Quantum non-linear sigma models: from quantum field theory to supersymmetry, conformal field theory, black holes, and strings / Sergei V. Ketov. Published/Created: Berlin; New York: Springer, 2000. Description: xii, 420 p.: ill.; 25 cm. ISBN:

3540674616 (alk. paper) Notes: Includes bibliographical references and index. Subjects: Sigma particles. Supersymmetry. Series: Texts and monographs in physics, 0172-5998 LC Classification: QC793.5.H42 K48 2000 Dewey Class No.: 530.14/3 21

Lethem, Jonathan. As she climbed across the table / Jonathan Lethem. Edition Information: 1st Vintage Contemporaries ed. Published/Created: New York: Vintage Books, 1998. Description: 212 p.: ill.; 21 cm. ISBN: 0375700129 Subjects: Physicists--Discoveries in science--Occupational neuroses--Black holes (Astronomy)--California--Fiction. Genre/Form: Humorous stories. Love stories. LC Classification: PS3562.E8544 A9 1998 Dewey Class No.: 813/.54 21

Lethem, Jonathan. As she climbed across the table / Jonathan Lethem. Edition Information: 1st ed. Published/Created: New York: Doubleday, 1997. Description: 212 p.; 22 cm. ISBN: 0385485174 Subjects: Physicists--Discoveries in science--Occupational neuroses--Black holes (Astronomy)--California--Fiction. Genre/Form: Love stories. Humorous stories. LC Classification: PS3562.E8544 A9 1997 Dewey Class No.: 813/.54 20

Miller, J. Hillis (Joseph Hillis), 1928- Black holes / J. Hillis Miller. J. Hillis Miller, or, Boustrophedonic reading / Manuel Asensi; translated by Mabel Richart. Published/Created: Stanford, Calif.: Stanford University Press, 1999. Related Names: Asensi, Manuel, 1959- J. Hillis Miller, or, Boustrophedonic reading. Description: xx, 537 p.; 24 cm. ISBN: 0804732434 (alk. paper) 0804732442 (pbk.: alk. paper) Notes: The Asensi portion of the book is a translation from an untitled and as yet unpublished manuscript in Spanish. Includes bibliographical references (p. [494]-537). Subjects: Trollope, Anthony, 1815-1882. Ayalás angel. Proust, Marcel, 1871-1922. A la recherche du temps perdu. Miller, J. Hillis (Joseph Hillis), 1928- -- Contributions in criticism. Criticism. Difference (Psychology) in literature. Series: Cultural memory in the present LC Classification: PN81 .M527 1999 Dewey Class No.: 801/.95 21

Miller, Jake. On the job with an astronomer: explorer of the universe / Jake Miller and Jonathan Rubinstein; illustrated by Susan Gal. Edition Information: 1st ed. Published/Created: Hauppauge, N.Y.: Barron's, 2001. Related Names: Rubinstein, Jonathan. Gal, Susan, ill. Description: 48 p.: ill.; 26 cm. ISBN: 0764118684 Summary: Explores the mysteries of deep space and the scientists whose job it is to search the night sky for comets, pulsars, and black holes, giving tips on how to determine if astronomy is the right career for you. Notes: Includes bibliographical references. Subjects: Astronomy--Vocational guidance--Juvenile literature. Astronomers. Vocational guidance. Occupations. Series: On the job with Bridgit & Hugo LC Classification: QB51.5 .M55 2001

Myers, Janet Nuzum, 1940- Strange stuff: true stories of odd places and things / by Janet Nuzum Myers; illustrations by Maj-Britt Hagsted. Published/Created: North Haven, Conn.: Linnet Books, 1999. Related Names: Hagsted, Maj-Britt. Description: viii, 104 p.: ill.; 23 cm. ISBN: 0208024050 (cloth/library: alk. paper) 0208024069 (paper: alk. paper) Summary: Kids love to contemplate strange stuff. Quicksand is right up there with Bigfoot, the Bermuda Triangle, meat=eating plants, the lost continent of Atlantis, wolf children, voodoo and zombies, scorpions and tarantulas, black holes, and more. The myths and the truth as the authors know them are here in this grab bag collection of weird stuff designed to pique the insatiable curiosity of children. And to get them thinking. Notes: Includes bibliographical references (p. 97-102) and index. Subjects: Science--Miscellanea--Juvenile literature. LC Classification: Q173 .M9826 1999 Dewey Class No.: 001.94 21

Newton, David E. Black holes and supernovae / David E. Newton. Edition Information: 1st ed. Published/Created: New York: Twenty-First Century Books, c1997. Description: 64 p.: ill. (some col.); 24 cm. ISBN: 0805044779 (alk. paper) Notes: Includes bibliographical references and index. Subjects: Black holes (Astronomy) Supernovae. Series: Secrets of space LC Classification: QB843.B55 N49 1997 Dewey Class No.: 523.8/875 21

Observational evidence for black holes in the universe: proceedings of a conference held in Calcutta, India, January 10-17, 1998 / edited by Sandip K. Chakrabarti. Published/Created: Dordrecht; Boston: Kluwer Academic Publishers, c1999. Related Names: Chakrabarti, Sandip K. Description: xi, 399 p.: ill.; 25 cm. ISBN: 079235298X (alk. paper) Notes: Includes bibliographical references and index. Subjects: Black holes (Astronomy)--Congresses. Series: Astrophysics and space science library; v. 234 LC Classification: QB843.B55 O28 1999 Dewey Class No.: 523.8/875 21

Oxlade, Chris. The mystery of black holes / Chris Oxlade. Published/Created: Des Plaines, Ill.: Heinemann Library, 1999. Description: 32 p.: col. ill.; 28 cm. ISBN: 1575728087(lib. bdg.) Summary: Explores the phenomenon of black holes, explains why astronomers think they exist, what causes them, what they are like inside, and the search to find black holes in space. Notes: Includes bibliographical references (p. 31) and index. Subjects: Black holes (Astronomy)--Juvenile literature. Black holes (Astronomy) Series: Can science solve? LC Classification: QB843.B55 O88 1999 Dewey Class No.: 523.8/875 21

Pacific Conference on Gravitation and Cosmology: February 1-6, 1996, Sheraton Walker-Hill, Seoul, Korea / edited by Y.M. Cho, C.H. Lee, S.W. Kim. Published/Created: Singapore; River Edge, N.J.: World Scientific, 1998. Related Names: Cho, Y. M. Lee, C. H. (Chul H.) Kim, S. W. Description: ix, 336 p.: ill.; 26 cm. ISBN: 9810236980 (alk. paper) Notes: Includes bibliographical references. Subjects: Astrophysics--Gravitation--Cosmology--Dark matter (Astronomy)--Black holes (Astronomy)--Congresses. LC Classification: QB460 .P33 1996 Dewey Class No.: 523.1´21

Parker, Steve. Space mysteries / Steve Parker. Published/Created: Austin, TX: Raintree Steck-Vaughn, c2002. Description: 32 p.: col. ill.; 28 cm. ISBN: 0739848534 Summary: Describes various outer space phenomena, including exploding stars, black holes, and pulsars. Notes: Includes index. Subjects: Astronomy--Juvenile literature. Space sciences--Juvenile literature. Astronomy. Outer space. Series: Space busters LC Classification: QB46 .P375 2002 Dewey Class No.: 520 21

Punsly, Brian, 1956- Black hole gravitohydromagnetics / Brian Punsly. Published/Created: Berlin; New York: Springer, c2001. Related Names: Punsly, Brian, 1956- Description: xii, 400 p.: ill. (some col.); 24 cm. ISBN: 3540414665 (alk. paper) Notes: Includes bibliographical references (p. 387-391) and index. Subjects: Black holes (Astronomy) Plasma astrophysics. Magnetohydrodynamics. Series: Astronomy and astrophysics library, 0941-7834 LC Classification: QB843.B55 P86 2001 Dewey Class No.: 523.8/875 21

Quantum gravity: proceedings of the sixth Moscow seminar: Moscow, Russia, June 12-19, 1995 / editors, V.A. Berezin, V.A. Rubakov, D.V. Semikoz. Published/Created: Singapore; River Edge, N.J.: World Scientific, c1998. Related Names: Berezin, V. A. Rubakov, V. A. Semikoz, D. V. Description: xxvii, 633 p.: ill.; 23 cm. ISBN: 9810230877 Notes: Includes bibliographical references. Subjects: Quantum gravity--Black holes (Astronomy)--Cosmology--Quantum Theory Black Holes (Astronomy)--Cosmology Conferences LC Classification: QC178 .S455 1995 Dewey Class No.: 530.14/3 21

Reeder, Jesse. Black holes and energy pirates: how to recognize and release them / Jesse Reeder. Published/Created: Freedom, CA: Crossing Press, 2001. Description: p. cm. ISBN: 1580910483 (pbk.: alk. paper) Subjects: Success--Psychological aspects. LC Classification: BF637.S8 R38 2001 Dewey Class No.: 158.1 21

Reviews in modern astronomy 14: dynamic stability and instabilities in the universe / Reinhard E. Schielicke, ed. Published/Created: Hamburg, Germany: Astronomische Gesellschaft, 2001. Related Names: Schielicke, Reinhard. Astronomische Gesellschaft (Germany). Meeting (2000: Bremen, Germany) Description: 332 p.: ill.; 24 cm. ISBN: 3980517640 Contents: Karl Schwarzschild Lecture: The Schwarzschild singularity - one clue to resolving the quantum measurement paradox / by Sir Roger Penrose -- Ludwig Biermann Award Lecture: The silent majority - jets and radio cores from low-luminosity black holes / by Heino Falcke -- Chaos in cosmos / by Peter H. Richter -- Chaos, comets, and the Kuiper Belt / by Martin J. Duncan, Harold Levison, Luke Dones, and Edward Thommes -- Planetary accretion - from planetesimals to protoplanet / by Eiichiro Kokubo -- Surprises from our sun / by Eric R. Priest. Large-scale structure - witness of evolution / by Dierck-Ekkehard Liebscher -- Dust induced structure formation / by Peter Woitke -- The FORS Deep Field / by Jochen Heidt [et al.] -- A map of the northern sky - the Sloan Digital Sky Survey in its first year / by Eva K. Grebel -- Mechanism and result of dynamical instabilities in hot stars / by Wolfgang Glatzel -- LBV nebulae - the mass lost from the most massive stars / by Kerstin Weis -- Dynamical evolution of star clusters / by Holger Baumgardt -- Warm and hot diffuse gas in dwarf galaxies / by Dominik J. Bomans. Notes: "Volume 14 continues the series with fourteen invited reviews and highlight contributions ... presented during the International Scientific Conference of the Society on 'dynamic stability and instabilities in the universe,' held at Bremen, September 18 to 23, 2000"--Preface. Includes bibliographical references and index. Subjects: Celestial mechanics--Congresses. Solar system--Congresses.

Ridpath, Ian. Facts on File stars & planets atlas / Ian Ridpath. Edition Information: Updated ed. Published/Created: New York: Facts on File, 2001. Related Names: Facts on File, Inc. Description: p. cm. ISBN: 0816048002 (alk. paper) Summary: An overview of the solar system, including such topics as the earth and other planets, the sun, moon, asteroids, comets, meteors, and black holes. Notes: Includes index. Subjects: Astronomy. Stars--Atlases. Planets--Atlases. Astronomy. Stars. Planets. LC Classification: QB45 .R53 2001 Dewey Class No.: 520 21

Ridpath, Ian. Facts on File stars & planets atlas / Ian Ridpath. Edition Information: 2nd ed. Published/Created: New York: Facts on File, 1997. Description: 80 p.: col. ill.; 30 cm. ISBN: 0816037167 Summary: An overview of the solar system, including such topics as the earth and other planets, the sun, moon, asteroids, comets, meteors, and black holes. Notes: "First published in Great Britain in 1992 by George Philip Limited." Includes index. Subjects: Astronomy. Stars--Atlases. Planets--Atlases. Astronomy. Stars. Planets. LC Classification: QB45 .R53 1997 Dewey Class No.: 520 21

Schilling, Govert. Flash!: the hunt for the biggest explosions in the universe / Govert Schilling; translated by Naomi Greenberg-Slovin. Published/Created: Cambridge: Cambridge University Press, 2002. Description: 291 p., [24] p. of plates: ill. (some col.), ports.; 24 cm. ISBN: 0521800536 Notes: Includes bibliographical references and index. Translated from the Dutch. Subjects: Gamma ray bursts. Black holes (Astronomy)

Segal, Justin. The amazing space almanac / by Justin Segal; illustrations by Carol Lyon.

Published/Created: Los Angeles: Lowell House Juvenile, c1998. Related Names: Lyon, Carol, 1963- ill. Description: 128 p.: ill. (some col.); 25 cm. ISBN: 1565656903 (hard) Summary: Discusses the many aspects of space, including the origin and nature of the universe, the history of space travel, quarks, quasars, black holes, and extraterrestrials. Notes: "A Roxbury Park book." Includes bibliographical references (p. 120-122) and index. Subjects: Astronomy--Juvenile literature. Astronomy--Miscellanea--Juvenile literature. Astronomy. Astronomy--Miscellanea. LC Classification: QB46 .S35 1998 Dewey Class No.: 520 21

Sipiera, Paul P. Black holes / by Paul P. Sipiera. Published/Created: New York: Children's Press, c1997. Description: 47 p.: ill. (some col.); 22 cm. ISBN: 0516203266 Summary: Explains the nature and formation of the cosmic phenomenon known as a black hole. Notes: Includes bibliographical references (p. 44) and index. Subjects: Black holes (Astronomy)--Juvenile literature. Black holes (Astronomy) Series: A true book LC Classification: QB843.B55 S57 1997 Dewey Class No.: 523.8/875 20

Strathern, Paul, 1940- Hawking and black holes / Paul Strathern. Edition Information: 1st Anchor Books ed. Published/Created: New York: Anchor Books, [1998] Description: 108 p.; 21 cm. ISBN: 0385492421 (pbk.) Notes: Includes bibliographical references (p. 107-108). Subjects: Hawking, S. W. (Stephen W.) Black holes (Astronomy) Physicists--Great Britain--Biography. Series: Strathern, Paul, 1940- Big idea. LC Classification: QC16.H33 S77 1998 Dewey Class No.: 503/.092 B 21

Strings, branes, and gravity: TASI 99: Boulder, Colorado, USA, 31 May-25 June 1999 / editors, Jeffrey Harvey, Shamit Kachru, Eva Silverstein. Published/Created: Singapore; River Edge, NJ: World Scientific, c2001. Related Names: Harvey, Jeffrey. Kachru, Shamit, 1970- Silverstein, Eva, 1970- Description: vii, 933 p., [3] leaves of plates: ill.; 23 cm. ISBN: 9810247745 Contents: TASI lectures on branes, black holes and anti-De Sitter space / M.J. Duff -- D-brane primer / Clifford V. Johnson -- TASI lectures on black holes in string theory / Amanda W. Peet -- TASI lectures: cosmology for string theorists / Sean M. Carroll -- TASI lectures on matrix theory / Tom Banks -- TASI lectures M theory phenomenology / Michael Dine -- TASI lectures: introduction to the AdS/CFT correspondence / Igor R. Klebanov -- TASI lectures on compactification and duality / David R. Morrison -- Compactification, Geometry and duality: N=2 / Paul S. Aspinwall -- TASI lectures on non-BPS D-brane systems / John H. Schwarz -- Lectures on warped compactifications and stringy brane constructions / Shamit Kachru -- TASI lectures on the holographic principle / Daniela Bigatti and Leonard Susskind. Notes: Includes bibliographical references. Subjects: String models--Gravity--Compactifications--Duality (Nuclear physics)--Black holes (Astronomy)--Supersymmetry--Branes--Congresses. LC Classification: QC794.6.S85 T48 1999 Dewey Class No.: 530.14 21

Superstrings and related matters: proceedings of the 1999 Spring Workshop on, The Abdus Salam ICTP, Trieste, Italy, 22-30 March 1999 / B. Greene ... [et al.]. Published/Created: New Jersey: World Scientific, c2000. Related Names: Greene, B. (Brian), 1963- Description: vii, 312 p.: ill.; 23 cm. ISBN: 9810241372 (alk. paper) Subjects: Superstring theories--Congresses. Black holes (Astronomy)--Congresses. LC Classification: QC794.6.S85 S65 1999 Dewey Class No.: 539.7/258 21

Superstrings and related matters: proceedings of the Trieste 2000 Spring Workshop on, ICTP, Trieste, Italy, 27 March-4 April 2000 / editors, C. Bachas ... [et al.]. Published/Created: Singapore; River

Edge, NJ: World Scientific, c2001. Related Names: Bachas, C. Description: vii, 205 p.: ill.; 22 cm. ISBN: 9810245254 Notes: Includes bibliographical references. Subjects: Superstring theories--Congresses. Black holes (Astronomy)--Congresses. LC Classification: QC794.6.S85 S65 2000 Dewey Class No.: 539.7/258 21

Taylor, Edwin F. Exploring black holes: introduction to general relativity / Edwin F. Taylor, John Archibald Wheeler. Published/Created: San Francisco: Addison Wesley Longman, 2000. Related Names: Wheeler, John Archibald, 1911- Description: 1 v. in various paging: ill.; 27 cm. ISBN: 020138423X Notes: Includes bibliographical references and index. Subjects: General relativity (Physics) Black holes (Astronomy) LC Classification: QC173.6 .T39 2000 Dewey Class No.: 530.11 21

Taylor, John Gerald, 1931- Black holes: the end of the universe? / John Taylor. Edition Information: Rev. and expanded ed. Published/Created: [London]: Souvenir Press, 1998. Description: viii, 197 p.; 22 cm. ISBN: 0285634437 (pbk.) Notes: Includes bibliographical references (p. [190]-193) and indexes. Subjects: Black holes (Astronomy)

The black hole: 25 years after / editors, Claudio Teitelboim, Jorge Zanelli. Published/Created: Singapore; River Edge, N.J.: World Scientific, c1998. Related Names: Zanelli, Jorge. Teitelboim, Claudio. Description: viii, 326 p.: ill.; 23 cm. ISBN: 9810233418 Notes: Includes bibliographical references. Subjects: Black holes (Astronomy) LC Classification: QB843.B55 B572 1998 Dewey Class No.: 523.8/875 21

The central regions of the galaxy and galaxies: proceedings of the 184th Symposium of the International Astronomical Union, held in Tokyo, Japan, August 18-22 / edited by Yoshiaki Sofue. Published/Created: Boston, Mass: Kluwer Academic, 1998. Related Names: Sofue, Yoshiaki. Description: xix, 521 p.: ill.; 25 cm. ISBN: 079235060X (hardbound: alk. paper) Notes: Includes bibliographical references and index. Subjects: Galaxies--Congresses. Stars--Clusters--Congresses. Stars--Formation--Congresses. Black holes (Astronomy)--Congresses. Galactic center--Congresses. Milky Way--Congresses. LC Classification: QB857.72 .I57 1998 Dewey Class No.: 523.1/12 21

The knowledge network / edited by Bibby Whittaker, Brian Smart. Published/Created: Brookfield, Conn.: Copper Beech Books, 1998. Related Names: Whittaker, Bibby. Smart, Brian. Description: p. cm. ISBN: 0761307796 Summary: A collection of interesting facts about topics such as fish courtship, snowboarding trickery, black holes, and the genius of Bach. Notes: Companion to: The knowledge factory. Includes index. Subjects: Handbooks, vade-mecums, etc.--Juvenile literature. Curiosities and wonders. LC Classification: AG105 .K635 1998 Dewey Class No.: 031.02 21

The neutron star-black hole connection / edited by Chryssa Kouveliotou, Joseph Ventura, and Ed van den Heuvel. Published/Created: Dordrecht; Boston: Kluwer Academic Publishers, c2001. Related Names: Kouveliotou, Chryssa. Ventura, Joseph. Heuvel, Edward Peter Jacobus van den, 1940- Description: xvii, 522 p.: ill.; 25 cm. ISBN: 1402002041 (alk. paper) Notes: Includes bibliographical references and indexes. Subjects: Neutron stars--Congresses. Black holes (Astronomy)--Congresses. Series: NATO science series. Series C, Mathematical and physical sciences; v. 567 LC Classification: QB843.N4 N379 1999 Dewey Class No.: 523.8/874 21

The Reader's Digest children's atlas of the universe / by Robert Burnham. Published/Created: Pleasantville, N.Y.: Reader's Digest Children's Pub., 2000. Related Names: Burnham, Robert, 1947- Weldon Owen Pty Limited. Description: 1 atlas (128 p.): col. ill., col. maps 35 cm. ISBN: 1575843730 (hardcover: alk.

paper) 157584379X (lib. ed.: alk. paper) Scale Information: Scale not given. Summary: Includes star charts, maps, and satellite photos of the planets and coverage of such topics as distant galaxies and black holes as well as hands-on projects. Notes: "Copyright 2000 Weldon Owen Pty Limited"--Prelim. p. [2]. Includes index. Subjects: Astronomy--Charts, diagrams, etc. Planets--Maps for children. Children's atlases. Universe. Astronomy. Planets. Atlases. Solar system--Maps for children. LC Classification: G1000 .R563 2000 Dewey Class No.: 523.2/022/3 21

The Scientific American book of the cosmos / David H. Levy, editor. Edition Information: 1st ed. Published/Created: New York: St. Martin's Press, c2000. Related Names: Levy, David H., 1948- Description: viii, 408 p., [16] p. of plates: ill. (some col.); 26 cm. ISBN: 0312254539 Contents: Preface / by D.H. Levy -- Some pieces from history. The heavens declare the glory of God / by A.G. Ingalls -- On the generalized theory of gravitation / by A. Einstein -- What is matter? / by E. Schrödinger -- The birth of the universe. The evolution of the universe / by P.J.E. Peebles ... [et al.] -- The evolution of galaxy clusters / by J.P. Henry, U.G. Briel, H. Böhringer -- The lives of quasars / by G. Cecil, J. Bland-Hawthorn, S. Veilleux -- Edwin Hubble and the expanding universe / by D.E. Osterbrock, J.A. Gwinn, R.S. Brashear -- The big bang and beyond: a century of modern cosmology / by L. Bernstein -- Galaxy formation. Galaxies behind the Milky Way / by R.C. Kraan-Korteweg, O. Lahav -- Colossal galactic explosions / by S. Veilleux, G. Cecil, J. Bland-Hawthorn -- Dark matter in the universe / by V. Rubin -- What's new in the Milky Way / by M.V. Magee -- The Milky Way. How the Milky Way formed / by S. van den Bergh, J.E. Hesser -- Black holes and the information paradox / by L. Susskind -- Collapse and formation of stars / by A.P. Boss -- Accretion disks in interacting binary stars / by J.K. Cannizzo, R.H. Kaitchuck -- The genesis of the solar system. The origins of the asteroids / by R.P. Binzel, M.A. Barucci, M. Fulchignoni -- The Kuiper belt / by J.X. Luu, D.C. Jewitt -- Migrating planets / by R. Malhotra -- Comet Shoemaker-Levy 9 meets Jupiter / by D.H. Levy, E. Shoemaker, C.S. Shoemaker -- On comets / by J.V. Scotti -- Solar system origins / by M. Washburn -- The planetary tour. Mercury: the forgotten planet / by R.M. Nelson -- Global climate change on Venus / by M.A. Bullock, D.H. Grinspoon -- The earth's elements / by R.P. Kirschner -- The evolution of the earth / by C.J. Allègre, S.H. Schneider -- Global climatic change on Mars / by J.S. Kargel, R.G. Strom -- The outer planets; The outer moons / M. Washburn -- What's new in the solar system / by P. Jedicke -- Life on earth ... and elsewhere. The structure of hereditary material / by F.H.C. Crick -- Life's far-flung raw materials / by M.P. Bernstein, S.A. Sandford, L.J. Allamandola -- The evolution of life on the earth / by S.J. Gould -- The case for relic life on Mars / by E.K. Gibson ... [et al.] -- Searching for life on other planets / by J. Roger, P. Angel, N.J. Woolf -- The search for extraterrestrial intelligence / by C. Sagan, F. Drake -- Giant planets orbiting faraway stars / by G.W. Marcy, R.P. Butler -- End of the proterozic eon / by A.H. Knoll -- The microverse. The discovery of the top quark / by T.M. Liss, P.L. Tipton -- Quantum philosophy / by J. Horgan -- Atoms, molecules, and the early universe / by A. Scheeline -- Speculation on endings and beginnings. The inflationary universe / by A.H. Guth, P.J. Steinhardt -- The expansion rate and size of the universe / by W.L. Freedman -- The self-reproducing inflationary universe / by A. Linde. Notes: "A Byron Preiss book." Includes bibliographical references (p. 401-406). Subjects: Cosmology. Astronomy. LC Classification: QB981 .S395 2000 Dewey Class No.: 523.1 21

The sixth Canadian Conference on General Relativity and Relativistic Astrophysics / Stephen P. Braham, Jack D. Gegenberg, Robert J. McKellar, editors. Published/Created: Providence, R.I.:

American Mathematical Society, c1997. Related Names: Braham, Stephen P. (Stephen Paul), 1964- Gegenberg, Jack D. (Jack David), 1947- McKellar, Robert J. (Robert James), 1952- Description: xiii, 373 p.; 26 cm. ISBN: 0821805231 Notes: "The sixth CCGRRA, held at the University of New Brunswick in Fredericton, New Brunswick on May 25-27, 1995"--Pref. Includes bibliographical references. Subjects: Black holes (Astronomy)--Congresses. General relativity (Physics)--Congresses. Relativistic astrophysics--Congresses. Series: Fields Institute communications; v. 15. LC Classification: QB843.B55 C34 1995 Dewey Class No.: 523.01 21

The universe. Published/Created: Alexandria, VA: Time-Life Books, c1998. Related Names: Time-Life Books. Description: 127 p.: ill. (some col.); 29 cm. ISBN: 0783513542 Summary: Examines the origin, structure, and workings of the universe, including galaxies, stars, dark matter, light years, black holes, and other aspects, and describes space exploration from ancient astronomy to modern probes. Notes: Includes index. Subjects: Cosmology--Astronomy--Juvenile literature. Universe. Series: Time-Life student library LC Classification: QB983 .U55 1998 Dewey Class No.: 520 21

Theory of black hole accretion disks / edited by Marek A. Abramowicz, Gunnlaugur Björnsson, James E. Pringle. Published/Created: Cambridge, UK; New York: Cambridge University Press, 1998. Related Names: Abramowicz, M. Gunnlaugur Björnsson Pringle, J. E. (James Edward), 1949- Description: xi, 293 p.: ill.; 26 cm. ISBN: 0521623626 Notes: Includes bibliographical references. Subjects: Black holes (Astronomy) Accretion (Astrophysics) Series: Cambridge contemporary astrophysics LC Classification: QB843.B55 T48 1998 Dewey Class No.: 523.8/875 21

This book really sucks!: the science behind gravity flight, leeches, black holes, tornadoes, our friend the vaccum cleaner, and most everything else that sucks / by the editors of Planet Dexter. Published/Created: New York: Planet Dexter, 1999. Related Names: Planet Dexter (Firm) Description: 80 p.: col. ill.; 24 cm. ISBN: 044844075X Summary: A humorous, but factual, discussion of all sorts of things that suck: gravity, leeches, babies, straws, and more. Notes: Rubber mat with suction cups covers entire book. Subjects: Vacuum--Juvenile literature. Science--Miscellanea. LC Classification: QC166 .B66 1999 Dewey Class No.: 533/.5 21

Vogt, Gregory. Stars / by Gregory L. Vogt. Published/Created: Mankato, Minn.: Bridgestone Books, c2002. Description: 24 p.: col. ill.; 21 cm. ISBN: 0736811214 Summary: Explores how stars form and create energy, including descriptions of black holes, constellations, and the movement of stars. Notes: Includes bibliographical references (p. 24) and index. Subjects: Stars--Juvenile literature. Stars. Series: Galaxy LC Classification: QB801.7 .V63 2002 Dewey Class No.: 523.8 21

Wheeler, John Archibald, 1911- Geons, black holes, and quantum foam: a life in physics / John Archibald Wheeler with Kenneth Ford. Edition Information: 1st ed. Published/Created: New York: Norton, c1998. Related Names: Ford, Kenneth William, 1926- Description: 380 p.: ill.; 25 cm. ISBN: 0393046427 Notes: Includes index. Subjects: Wheeler, John Archibald, 1911- Physics--History. Astronomy--History. Physicists--United States--Biography. LC Classification: QC16.W48 A3 1998 Dewey Class No.: 530/.092 B 21

Wilson, Colin, 1931- Mysteries of the universe / Colin Wilson. Edition Information: 1st American ed. Published/Created: New York: DK Pub., 1997. Description: 37 p.: ill. (some col.), col. maps; 29 cm. ISBN: 0789421658 Summary: Explores such mysteries as Bigfoot, the Loch Ness Monster, weird weather, and black holes.

Notes: Includes index. Subjects: Science--Curiosities and wonders--Miscellanea--Juvenile literature. Curiosities and wonders--Miscellanea. Series: Unexplained (New York, N.Y.) LC Classification: Q163 .W55 1997 Dewey Class No.: 001.9 21

Wolfson, Richard. Einstein's relativity & the quantum revolution [sound recording] / [Richard Wolfson]. Published/Created: Springfield, VA: Teaching Co., [1999] Related Names: Teaching Company. Description: 12 sound discs: analog, Dolby processed + 2 course guides (22 cm.) Publisher Number: 151A Teaching Co. 152A Teaching Co. Contents: pt. 1: lecture 1. Time travel, tunneling, tennis, and tea; lecture 2. Heaven and earth, place and motion; lecture 3. The clockwork universe; lecture 4. Let there be light!; lecture 5. Speed c relative to what?; lecture 6. Earth and the ether: a crisis in physics; lecture 7. Einstein to the rescue; lecture 8. Uncommon sense: stretching time; lecture 9. Muons and time-traveling twins; lecture 10. Escaping contradiction: simultaneity is relative; lecture 11. Faster than light? Past, future, and elsewhere; lecture 12. What about $E=mc^2$, and is everything relative? -- pt. 2: lecture 13. A problem of gravity; lecture 14. Curved spacetime; lecture 15. Black holes; lecture 16. Into the heart of matter; lecture 17. Enter the quantum; lecture 18. Wave or particle?; lecture 19. Quantum uncertainty: farewell to determinism; lecture 20. Particle or wave?; lecture 21. Quantum weirdness and Schrödinger's cat; lecture 22. The particle zoo; lecture 23. Cosmic connections; lecture 24. Toward a theory of everything. Notes: Teaching Co.: 151A--152A. In 2 containers (26 cm.). Cast: Richard Wolfson, lecturer. Subjects: Einstein, Albert, 1879-1955. Relativity (Physics) Space and time. Quantum theory. Series: The great courses on tape LC Classification: RZA 6403, etc.

Zahn, Timothy. Angelmass / Timothy Zahn. Edition Information: 1st ed. New York: Tor, 2001. Description: 430 p.; 24 cm. ISBN: 0312878281 (alk. paper) Notes: "A Tom Doherty Associates book." Subjects: Black holes (Astronomy)--Fiction. Space colonies--Fiction. Genre/Form: Science fiction. LC Classification: PS3576.A33 A8 2001 Dewey Class No.: 813/.54 21

JOURNALS AND MAGAZINES

Abdalla, Elcio; Correa-Borbonet, L. Alejandro Black Hole Entropy by the Brick-Wall Method in Four and Five Dimensions with U(1) Charges Journal Modern Physics Letters A Date: Dec 01 2001 Abstract: Using the brick-wall method the authors compute the statistical entropy of a scalar field in a nontrivial background, in two different cases. These backgrounds are generated by four- and five-dimensional black holes with four and three U(1) charges respectively. The Bekenstein entropy formula is generally obeyed, but corrections are discussed in the latter case. Journal volume: 16 Issue number: 39 Rights: World Scientific Publishing Company Publisher: World Scientific Publishing Company

Abramowicz, Marek A.; Percival, Marcus J. Thermal equilibria of accretion discs around black holes: a review of the recent progress Journal Classical and Quantum Gravity Date: Aug 01 1997 Abstract: Accretion discs around supermassive black holes provide a model explaining observed properties of quasars and other active galactic nuclei. Theory of accretion discs now undergoes substantial progress and changes. This review describes new understanding of thermal equilibria of black hole accretion discs. Journal volume: 14 Issue number: 8 Page number: 2003-2017 Publisher: Institute of Physics

Achucarro, A; Achucarro, A; Achucarro, A.; Gregory, R. Selection Rules for Splitting Strings Journal Physical Review Letters Date: Sep 30 1997 Abstract: It has been pointed out that Nielsen-Olesen vortices may be able to decay by pair production of black holes. The authors show that when the Abelian-Higgs model is embedded in a larger theory, the additional fields may lead to selection rules for this process even in the absence of fermions due to the failure of a charge quantization condition. Journal volume: 79 Issue number: 11 Page number: 1972-1975 Subjects: String Models/selection rules; Black Holes/Pair Production; Vortices; Higgs Model; Quantization; Topology; Defects; Monopoles; Fermions; Energy-Momentum Tensor

Adams, F.C.; Laughlin, G. A dying universe: the long-term fate and evolution of astrophysical objects Journal Reviews of Modern Physics Date: Apr 30 1997 Abstract: Astrophysical issues related to the long-term fate of the universe are outlined. The evolution of planets, stars, stellar populations, galaxies, and the universe itself over time scales that greatly exceed the current age of the universe are considered. Their discussion starts with new stellar evolution calculations which follow the future evolution of the low-mass (M-type) stars that dominate the stellar mass function. They derive scaling relations that describe how the range of stellar masses and lifetimes depends on forthcoming increases in metallicity. Journal volume: 692 Page number: 337-372 Subjects: Galactic Evolution; Star Evolution; Astronomy; Galaxies; Mass;

Neutron Stars; Planets; Populations; Reviews; Scaling; White Dwarf Stars; Document Types; Dwarf Stars; Stars

Adams, Fred C.; Kane, Gordon L.; Mbonye, Manasse; Perry, Malcolm J. Proton Decay, Black Holes, and Large Extra Dimensions Journal International Journal of Modern Physics A Date: May 01 2001 Abstract: The authors consider proton decay in theories that contain large extradimensions. If virtual black hole states are allowed by the theory, asis generally the case, then proton decay can proceed via virtual blackholes. The experimental limits on the proton lifetime place strong constraints on the quantum gravity scale M qg (the effective Planck mass). For most theories, this implies a lower boundof M qg >1016 GeV. The corresponding bound on the size of large extradimensions is ℓ<106/n×10-30 cm,where n is the number of such dimensions.

Agnese, A. G.; La Camera, M. Kerr-AdS and Kerr-dS solutions reexamined Journal Physical Review Date: Apr 15 2000 Abstract: Abstract available at publisher web site Journal volume: 61 Issue number: 8 Page number: 087502-087502-4 Publisher: American Physical Society

Akhmedov, E. T. Black Hole Thermodynamics from the Point of View of Superstring Theory Journal International Journal of Modern Physics A Date: Jan 10 2000 Abstract: In this review the authors try to give a pedagogical introduction to the recent progress in the resolution of old problems of black hole thermodynamics within superstring theory. The authors start with a brief description of classical black hole dynamics. Then, follow with the consideration of general properties of supersymmetric black holes. The authors conclude with the review of the statistical explanation of the black hole entropy and string theory description of the black hole evaporation. Journal volume: 15 Issue number: 01 Rights: World Scientific Publishing Company Publisher: World Scientific Publishing Company

Alcubierre, Miguel; Allen, Gabrielle; Brügmann, Bernd; Lanfermann, Gerd; Seidel, Edward; Suen, Wai-Mo; Tobias, Malcolm Gravitational collapse of gravitational waves in 3D numerical relativity Journal Physical Review Date: Feb 15 2000 Abstract: Abstract available at publisher web site Journal volume: 61 Issue number: 4 Page number: 041501-041501-5 Publisher: American Physical Society

Alcubierre, Miguel; Benger, Werner; Brügmann, Bernd; Lanfermann, Gerd; Nerger, Lars; Seidel, Edward; Takahashi, Ryoji 3D Grazing Collision of Two Black Holes Journal Physical Review Letters Date: Dec 31 2001 Abstract: Abstract available at publisher web site Journal volume: 87 Issue number: 27 Page number: 271103-271103-4 Publisher: American Physical Society

Alcubierre, Miguel; Brügmann, Bernd; Dramlitsch, Thomas; Font, José A.; Papadopoulos, Philippos; Seidel, Edward; Stergioulas, Nikolaos; Takahashi, Ryoji Towards a stable numerical evolution of strongly gravitating systems in general relativity: The conformal treatments Journal Physical Review Date: Aug 15 2000 Abstract: Abstract available at publisher web site Journal volume: 62 Issue number: 4 Page number: 044034-044034-16 Publisher: American Physical Society

Alcubierre, Miguel; Brügmann, Bernd; Pollney, Denis; Seidel, Edward; Takahashi, Ryoji Black hole excision for dynamic black holes Journal Physical Review Date: Sep 15 2001 Abstract: Abstract available at publisher web site Journal volume: 64 Issue number: 6 Page number: 061501-061501-5 Publisher: American Physical Society

Alexeyev, S. O.; Sazhin, M. V.; Pomazanov, M. V.; Starobinsky, A. A. Black Holes of a Minimal Size in String Gravity Journal International Journal of Modern Physics Date: Apr 01 2001 Abstract: A lower limit for a neutral black hole size is obtained in the frames of the string gravity model with the second order curvature correction. It is shown that this effect remains when the third order curvature correction is also taken into account and argued that such restriction does exist in all perturbative orders of curvature expansions. Journal volume: 10 Issue number: 02 Rights: World Scientific Publishing Company Publisher: World Scientific Publishing Company

Alexeyev, S.; Mignemi, S. Black holes and naked singularities in the low-energy limit of string gravity with a modulus field Journal Classical and Quantum Gravity Date: Oct 21 2001 Abstract: The authors show that the black hole solutions of the effective string theory action, where one-loop effects that couple the moduli to gravity via a Gauss-Bonnet term are taken into account, admit primary scalar hair. The requirement of absence of naked singularities imposes an upper bound on the scalar charges. Journal volume: 18 Issue number: 20 Page number: 4165-4177 Publisher: Institute of Physics

Alvi, Kashif Approximate binary-black-hole metric Journal Physical Review Date: Jun 15 2000 Abstract: Abstract available at publisher web site Journal volume: 61 Issue number: 12 Page number: 124013-124013-19 Publisher: American Physical Society

Alvi, Kashif Energy and angular momentum flow into a black hole in a binary Journal Physical Review Date: Nov 15 2001 Abstract: Abstract available at publisher web site Journal volume: 64 Issue number: 10 Page number: 104020-104020-9 Publisher: American Physical Society

Åminneborg, Stefan; Bengtsson, Ingemar; Brill, Dieter; Holst, Sören; Peldán, Peter Black holes and wormholes in dimensions Journal Classical and Quantum Gravity Date: Mar 01 1998 Abstract: A large variety of spacetimes - including the BTZ black holes - can be obtained by identifying points in (2 + 1)-dimensional anti-de Sitter space by means of a discrete group of isometries. The authors consider all such spacetimes that can be obtained under a restriction to time-symmetric initial data and one asymptotic region only. The resulting spacetimes are non-eternal black holes with collapsing wormhole topologies. The authors' approach is geometrical, and they discuss in detail the allowed topologies, the shape of the event horizons, topological censorship and trapped curves. Journal volume: 15 Issue number: 3 Page number: 627-644 Publisher: Institute of Physics

Anderson, Paul R.; Hiscock, William A.; Taylor, Brett E. Do Semiclassical Zero Temperature Black Holes Exist? Journal Physical Review Letters Date: Sep 18 2000 Abstract: Abstract available at publisher web site Journal volume: 85 Issue number: 12 Page number: 2438-2441 Publisher: American Physical Society

Andersson, Nils; Glampedakis, Kostas Super radiance Resonance Cavity Outside Rapidly Rotating Black Holes Journal Physical Review Letters Date: May 15 2000 Abstract: Abstract available at publisher web site Journal volume: 84 Issue number: 20 Page number: 4537-4540 Publisher: American Physical Society

Anninos, Peter; Rothman, Tony Instability of extremal relativistic charged spheres Journal Physical Review Date: Jan 15 2002 Abstract: Abstract available at publisher web site Journal volume: 65 Issue number: 2 Page number: 024003-024003-10 Publisher: American Physical Society

Araújo, M.E.; Letelier, P.S.; Oliveira, S.R. Two Kerr black holes with axisymmetric spins: head-on collision and gravitational radiation Journal Classical and Quantum Gravity Date: Oct 01 1998 Abstract: The authors present a semi-analytical approach to the interaction of two (originally) Kerr black holes through a head-on collision process. An expression for the rate of emission of gravitational radiation is derived from an exact solution to Einstein's field equations. The total amount of gravitational radiation emitted in the process is calculated. They find that the spin-spin interaction increases the emission of gravitational wave energy by up to 0.2% of the total rest mass. They also discuss the possibility of spin exchange between the holes. Journal volume: 15 Issue number: 10 Page number: 3051-3060 Publisher: Institute of Physics

Aros, Rodrigo; Troncoso, Ricardo; Zanelli, Jorge Black holes with topologically nontrivial AdS asymptotics Journal Physical Review Date: Apr 15 2001 Abstract: Abstract available at publisher web site Journal volume: 63 Issue number: 8 Page number: 084015-084015-12 Publisher: American Physical Society

Ashtekar, Abhay; Beetle, Christopher; Lewandowski, Jerzy Mechanics of rotating isolated horizons Journal Physical Review Date: Aug 15 2001 Abstract: Black hole mechanics was recently extended by replacing the more commonly used event horizons in stationary space-times with isolated horizons in more general space-times (which may admit radiation arbitrarily close to black holes). However, so far the detailed analysis has been restricted to nonrotating black holes (although it incorporated arbitrary distortion, as well as electromagnetic, Yang-Mills, and dilatonic charges).

Ashtekar, Abhay; Corichi, .Alejandro Laws governing isolated horizons: inclusion of dilaton couplings Journal Classical and Quantum Gravity Date: Mar 21 2000 Abstract: Mechanics of non-rotating black holes was recently generalized by replacing the static event horizons used in standard treatments with 'isolated horizons'. This framework is extended to incorporate dilaton couplings. Since there can be gravitational and matter radiation outside isolated horizons, now the fundamental parameters of the horizon, used in mechanics, must be defined using only the local structure of the horizon, without reference to infinity. Journal volume: 17 Issue number: 6 Page number: 1317-1332 Publisher: Institute of Physics

Ashtekar, Abhay; Corichi, Alejandro; Sudarsky, Daniel Hairy black holes, horizon mass and solitons Journal Classical and Quantum Gravity Date: Mar 07 2001 Abstract: P Properties of the horizon mass of hairy black holes are discussed with emphasis on certain subtle and initially unexpected features. A key property suggests that hairy black holes may be regarded as 'bound states' of ordinary black holes (without hair) and coloured solitons. This model is then used to predict the qualitative behaviour of the horizon properties of hairy black holes, to provide a physical 'explanation' of their instability and to put qualitative constraints on the end-point configurations that result from this instability. The available numerical calculations support these predictions. Journal volume: 18 Issue number: 5 Page number: 919-940 Publisher: Institute of Physics

Ashtekar, Abhay; Fairhurst, Stephen; Krishnan, Badri Isolated horizons: Hamiltonian evolution and the first law Journal Physical Review Date: Nov 15 2000 Abstract: Abstract available at publisher web site Journal volume: 62 Issue number: 10 Page number: 104025-104025-29 Publisher: American Physical Society

Ashworth, M. C.; Hayward, Sean A. Boundary terms and Noether current of spherical black holes Journal Physical Review Date: Oct 15 1999 Abstract: Abstract available at publisher web site Journal volume: 60 Issue number: 8 Page number: 084004-084004-8 Publisher: American Physical Society

Ashworth, M. C.; Hayward, Sean A. Noether currents of charged spherical black holes Journal Physical Review Date: Sep 15 2000 Abstract: Abstract available at publisher web site Journal volume: 62 Issue number: 6 Page number: 064024-064024-3 Publisher: American Physical Society

Augusto, Pedro; Wilkinson, Peter N. An observational constraint on gravitational lensing by objects of mass 109.5-1010.9 M

Journal Monthly Notices of the Royal Astronomical Society Date: Jan 01 2001 Abstract: A radio-based search for strong gravitational lensing with image separations in the range 160-300 milliarcsec (mas) has yielded a null result for a sample of 1665 sources the mean redshift of which is estimated to be ~1.3. The lensing rate for this previously unexplored separation range, <1:555 at the 95 per cent confidence level, is less than on arcsecond-scales - as expected from models of lensing galaxy populations. Lensing on 160-300 mas scales is expected to arise predominantly from spiral galaxies at a rate dependent on the disk-halo mass ratio and the evolving number density of the population with redshift.

Awad, Adel M.; Johnson, Clifford V. Higher dimensional Kerr-AdS black holes and the AdS/CFT correspondence Journal Physical Review Date: Jun 15 2001 Abstract: Using the counterterm subtraction technique the authors calculate the stress-energy tensor, action, and other physical quantities for Kerr-AdS black holes in various dimensions. For Kerr-AdS$_5$ black holes with both rotation parameters nonzero, the authors demonstrate that the stress-energy tensor, in the zero mass parameter limit, is equal to the stress tensor of the weakly coupled four dimensional dual field theory. As a result, the total energy of the general Kerr-AdS$_5$ black hole at zero mass parameter exactly matches the Casimir energy of the dual field theory.

Awad, Adel M.; Johnson, Clifford V. Higher dimensional Kerr-AdS black holes and the AdS/CFT correspondence Journal Physical Review Date: Jun 15 2001 Abstract: Abstract available at publisher web site Journal volume: 63 Issue number: 12 Page number: 124023-124023-9 Publisher: American Physical Society

Awad, Adel M.; Johnson, Clifford V. Holographic stress tensors for Kerr-AdS black holes Journal Physical Review Date: Apr 15 2000 Abstract: Abstract available at publisher web site tcal: N=4 supersymmetric Yang-Mills theory on the rotating Einstein universe, the authors explicitly verify the equality of the zero mass stress tensor from the two sides of the correspondence, and present the result for a general mass as a prediction from gravity. Amusingly, it is observed in four dimensions that while the trace of the stress tensor defined using the standard counter terms does not vanish, its integral does, thereby keeping the action free of ultraviolet divergences. Journal volume: 61 Issue number: 8 Page number: 084025-084025-9 Publisher: American Physical Society

Awad, Adel M.; Johnson, Clifford V. Scale versus conformal invariance in the AdS/CFT correspondence Journal Physical Review Date: Dec 15 2000 Abstract: Abstract available at publisher web site Journal volume: 62 Issue number: 12 Page number: 125010-125010-4 Publisher: American Physical Society

Ayón-Beato, Eloy "No-hair" theorem for spontaneously broken Abelian models in static black holes Journal Physical Review Date: Nov 15 2000 Abstract: Abstract available at publisher web site Journal volume: 62 Issue number: 10 Page number: 104004-104004-7 Publisher: American Physical Society

Ayón-Beato, Eloy; García, Alberto; Macías, Alfredo; Quevedo, Hernando Uniqueness theorems for static black holes in metric-affine gravity Journal Physical Review Date: Apr 15 2000 Abstract: Abstract available at publisher web site Journal volume: 61 Issue number: 8 Page number: 084017-084017-6 Publisher: American Physical Society

Ayón-Beato, Eloy; García, Alberto; Macías, Alfredo; Quevedo, Hernando Static black holes of metric-affine gravity in the presence of matter Journal Physical Review Date: Jul 15 2001 Abstract: Abstract available at publisher web site Journal volume: 64 Issue number: 2 Page number: 024026-024026-7 Publisher: American Physical Society

Azreg-Aïnou, Mustapha Instability of two-dimensional heterotic stringy black holes Journal Classical and Quantum Gravity Date: Jan 01 1999 Abstract: The authors solve the eigenvalue problem of general relativity for the case of charged black holes in two-dimensional heterotic string theory, derived by McGuigan et al. For the case of , the authors find a physically acceptable time-dependent growing mode; thus the black hole is unstable. The extremal case is stable. Journal volume: 16 Issue number: 1 Page number: 245-253 Publisher: Institute of Physics

Baaquie, Belal E.; Kwek, L. C. Superstrings, Gauge Fields and Black Holes Journal International Journal of Modern Physics A Date: Jun 01 2001 Abstract: There has been spectacular progress in the development of string and superstring theories since its inception thirty years ago. Development in this area has never been impeded by the lack of experimental confirmation. Indeed, numerous bold and imaginative strides have been taken and the sheer elegance and logical consistency of the argument shave served as a primary motivation for string theorists to push their formulations ahead. In fact the development in this area has been so rapid that new ideas quickly become obsolete. On the other hand, this rapid development has proved to be the greatest hindrance for novices interested in this area. These notes serve as a gentle introduction to this topic. Journal volume: 16 Issue number: 15 Rights: World Scientific Publishing Company Publisher: World Scientific Publishing Company

Baker, J.; Brügmann, B.; Campanelli, M.; Lousto, C. O.; Takahashi, R. Plunge Waveforms from Inspiralling Binary Black Holes Journal Physical Review Letters Date: Sep 17 2001 Abstract: Abstract available at publisher web site Journal volume: 87 Issue number: 12 Page number: 121103-121103-4 Publisher: American Physical Society

Baker, John; Brandt, Steven; Campanelli, Manuela; Lousto, Carlos O.; Seidel, Edward; Takahashi, Ryoji Nonlinear and perturbative evolution of distorted black holes: Odd-parity modes Journal Physical Review Date: Dec 15 2000 Abstract: Abstract available at publisher web site Journal volume: 62 Issue number: 12 Page number: 127701-127701-4 Publisher: American Physical Society

Baker, John; Li, Chun Biu The two-phase approximation for black hole collisions: is it robust? Journal Classical and Quantum Gravity Date: May 01 1997 Abstract: Recently, Abrahams and Cook devised a method of estimating the total radiated energy resulting from collisions of distant black holes by applying Newtonian evolution to the holes up to the point where a common apparent horizon forms around the two black holes and subsequently applying Schwarzschild perturbation techniques. Despite the crudeness of their method, their results for the case of head-on collisions were surprisingly accurate. Here the authors take advantage of the simple radiated energy formula devised in the close-slow approximation for black hole collisions to test how strongly the Abrahams - Cook result depends on the choice of moment when the method of evolution switches over from Newtonian to general relativistic evolution. The authors find that their result is robust, not depending strongly on this choice. Journal volume: 14 Issue number: 5 Page number: L77-L82 Publisher: Institute of Physics

Balasubramanian, V.; Leigh, R.G. D-branes, moduli, and supersymmetry Journal Physical Review, D (Particles Fields) Date: May 31 1997 Abstract: The authors study toroidal compactifications of type II string theory with D-branes and nontrivial antisymmetric tensor moduli and show that turning on these fields modifies the supersymmetry projections imposed by D-branes. These modifications are seen to be necessary for the consistency of T duality. The authors also show the existence of unusual BPS configurations of branes at angles that are supersymmetric because of conspiracies between moduli fields. Journal volume: 55 Issue number: 10 Page number: 6415-6422 Subjects: String Models/supersymmetry; String Models/compactification; Supersymmetry; Compactification; Tensors; Duality; Quantum Field Theory; Black Holes

Balasubramanian, V; Lawrence, A.; Kraus, P.; Trivedi, S.P. Holographic probes of anti–de Sitter spacetimes Journal Physical Review, D (Particles Fields) Date: May 31 1999 Abstract: The authors describe probes of anti–de Sitter spacetimes in terms of conformal field theories on the AdS boundary. Our basic tool is a formula that relates bulk and boundary states classical bulk field configurations are dual to expectation values of operators on the boundary. At the quantum level the authors relate the operator expansions of bulk and boundary fields. Journal volume: 59 Issue number: 10 Page number: 104021 Subjects: Space-Time; Gravitation; Duality; String Models; Conformal Invariance; Black Holes; Metrics

Balasubramanian, Vijay; de Boer, Jan; Keski-Vakkuri, Esko; Ross, Simon F. Supersymmetric conical defects: Towards a string theoretic description of black hole formation Journal Physical Review Date: Sep 15 2001 Abstract: Conical defects, or point particles, in AdS_3 are one of the simplest nontrivial gravitating systems, and are particularly interesting because black holes can form from their collision. The authors embed the BPS conical defects of three dimensions into the N=4b supergravity in six dimensions, which arises from the IIB string theory compactified on K3. The required Kaluza-Klein reduction of the six dimensional theory on a sphere is analyzed in detail, including the relation to the Chern-Simons supergravities in three dimensions.

Balasubramanian, Vijay; de Boer, Jan; Keski-Vakkuri, Esko; Ross, Simon F. Supersymmetric conical defects: Towards a string theoretic description of black hole formation Journal Physical Review Date: Sep 15 2001 Abstract: Abstract available at publisher web site tcal: N=4b supergravity in six dimensions, which arises from the IIB string theory compactified on K3. The required Kaluza-Klein reduction of the six dimensional theory on a sphere is analyzed in detail, including the relation to the Chern-Simons supergravities in three dimensions. The authors show that the six dimensional spaces obtained by embedding the 3D conical defects arise in the near-horizon limit of rotating black strings. Various properties of these solutions are analyzed and the authors propose a representation of our defects in the CFT dual to asymptotically AdS 3×S 3 spaces. Our work is intended as a first step towards analyzing colliding defects that form black holes. Journal volume: 64 Issue number: 6 Page number: 064011-064011-19 Publisher: American Physical Society

Balasubramanian, Vijay; Ross, Simon F. Holographic particle detection Journal Physical Review Date: Feb 15 2000 Abstract: Abstract available at publisher web site Journal volume: 61 Issue number: 4 Page number: 044007-044007-12 Publisher: American Physical Society

Balbinot, R.; Fabbri, A. Two-dimensional black holes in accelerated frames: quantum aspects Journal Classical and Quantum Gravity Date: Feb 01 1997 Abstract: By considering charged black-hole solutions of a one-parameter family of two-dimensional dilaton gravity theories, one finds the existence of quantum mechanically stable gravitational kinks with a simple mass-to-charge relation. Unlike their Einsteinian counterpart (i.e. extreme Reissner - Nordström), these have nonvanishing horizon surface gravity. Journal volume: 14 Issue number: 2 Page number: 463-476 Publisher: Institute of Physics

Ballantyne, D. R.; Iwasawa, K.; Fabian, A. C. Evidence for ionized accretion discs in five narrow-line Seyfert 1 galaxies Journal Monthly Notices of the Royal Astronomical Society Date: May 01 2001 Abstract: The authors present the results of fitting ASCA spectra of six narrow-line Seyfert 1 (NLS1) galaxies with the ionized reflection models of Ross & Fabian. The authors find that five of the galaxies (TON S 180, PKS 0558-504, Ark 564, Mrk 335 and PG 1244+026) are well fitted by the ionized disc model, and these are often better fits than the alternative models considered. The sixth galaxy, NGC 4051, has additional spectral complexity that cannot be well described by a simple ionized disc model or any of the other alternative models.

Banados, M. Dilaton gravity (with a Gauss-Bonnet term) derived from five-dimensional Chern-Simons gravity Journal Physical Review, D (Particles Fields) Date: Feb 28 1997 Abstract: The authors study the problem of boundary terms and boundary conditions for Chern-Simons gravity in five dimensions. The authors show that under reasonable boundary conditions one finds an effective field theory at the four-dimensional boundary described by dilaton gravity with a Gauss-Bonnet term. Journal volume: 55 Issue number: 4 Page number: 2051-2058 Subject Keyword: Quantum Gravity/Many-Dimensional Calculations; Gravitation; Boundary Conditions; Quantum Field Theory; Coupling; Matter; Conformal Invariance; Black Holes; Entropy

Bañados, Máximo Twisted sectors in three-dimensional gravity Journal Physical Review Date: Nov 15 1999 Abstract: Abstract available at publisher web site Journal volume: 60 Issue number: 10 Page number: 104022-104022-10 Publisher: American Physical Society

Bañados, Máximo; Gomberoff, Andrés; Martínez, Cristián Anti-de Sitter space and black holes Journal Classical and Quantum Gravity Date: Nov 01 1998 Abstract: Anti-de Sitter space with identified points give rise to black-hole structures. This was first pointed out in three dimensions and generalized to higher dimensions by Aminneborg et al. In this paper, the authors analyse several aspects of the five-dimensional anti-de Sitter black hole, including its relation to thermal anti-de Sitter space, its embedding in a Chern - Simons supergravity theory, its global charges and holonomies and the existence of Killing spinors. Journal volume: 15 Issue number: 11 Page number: 3575-3598 Publisher: Institute of Physics

Banks, T.; Fischler, W.; Klebanov, I.R.; Susskind, L. Schwarzschild Black Holes from Matrix Theory Journal Physical Review Letters Date: Jan 01 1998 Abstract: The authors consider matrix theory compactified on T^3 and show that it correctly describes the properties of Schwarzschild black holes in 7+1 dimensions, including the mass-entropy relation, the Hawking temperature, and the physical size, up to numerical factors of order unity. The most economical description involves setting the cutoff N in the discretized light-cone quantization to be of order the black hole entropy. A crucial ingredient necessary for our work is the recently proposed equation of state for 3+1 dimensional supersymmetric Yang-Mills theory with 16 supercharges. The authors give detailed arguments for the range of validity of this equation following the methods of Horowitz and Polchinski. Journal volume: 80 Issue number: 2 Page number: pp.226-229

Barack, Leor Late time decay of scalar, electromagnetic, and gravitational perturbations outside rotating black holes Journal Physical Review Date: Jan 15 2000 Abstract: Abstract available at publisher web site Journal volume: 61 Issue number: 2 Page number: 024026-024026-27 Publisher: American Physical Society

Barack, Leor; Ori, Amos Late-time decay of gravitational and electromagnetic perturbations along the event horizon Journal Physical Review Date: Dec 15 1999 Abstract: Abstract available at publisher web site Journal volume: 60 Issue number: 12 Page number: 124005-124005-16 Publisher: American Physical Society

Barack, Leor; Ori, Amos Late-Time Decay of Scalar Perturbations Outside Rotating

Black Holes Journal Physical Review Letters Date: May 31 1999 Abstract: Abstract available at publisher web site Journal volume: 82 Issue number: 22 Page number: 4388-4391 Publisher: American Physical Society

Bardeen, James; Horowitz, Gary T. Extreme Kerr throat geometry: A vacuum analog of AdS $2 \times S 2$ Journal Physical Review Date: Nov 15 1999 Abstract: Abstract available at publisher web site Journal volume: 60 Issue number: 10 Page number: 104030-104030-10 Publisher: American Physical Society

Bars, I. Black hole entropy reveals a twelfth dimension Journal Physical Review, D (Particles Fields) Date: Mar 31 1997 Abstract: The Bekenstein-Hawking black hole entropy in string theory and its generalizations, as expressed in terms of charges that correspond to central extensions of the supersymmetry algebra, has more symmetries than U duality. It is invariant under transformations of the charges, involving a twelfth (or thirteenth) "dimension." Journal volume: 55 Issue number: 6 Page number: 3633-3641 Subjects: String Models; Entropy; Many-Dimensional Calculations; Charges; Supersymmetry; Algebra; Duality

Barvinsky, Andrei; Das, Saurya; Kunstatter, Gabor Spectrum of charged black holes—the big fix mechanism revisited Journal Classical and Quantum Gravity Date: Nov 21 2001 Abstract: Following an earlier suggestion of the authors (Barvinsky A and Kunstatter G 1997 Mass spectrum for black holes in generic 2-D dilaton gravity Proc. 2nd International A D Sakharov Conference on Physics ed I M Dremin and A M Seminkhatov (Singapore: World Scientific) pp 210-15), the authors use some basic properties of Euclidean black hole thermodynamics and the quantum mechanics of systems with periodic phase space coordinate to derive the discrete two-parameter area spectrum of generic charged spherically symmetric black holes in any dimension.

Bassett, Bruce A.; Tsujikawa, Shinji Inflationary preheating and primordial black holes Journal Physical Review Date: Jun 15 2001 Abstract: Preheating after inflation may overproduce primordial black holes (PBH's) in many regions of parameter space. As an example the authors study two-field models with a massless self-interacting inflation, taking into account second order field and metric back reaction effects as spatial averages. The authors find that a complex quilt of parameter regions above the Gaussian PBH overproduction threshold emerges due to the enhancement of curvature perturbations on all scales. It should be possible to constrain realistic models of inflation through PBH overproduction although many issues, such as rescattering and non-Gaussianity, remain unsolved or unexplored. Journal volume: 63 Issue number: 12 Page number: 123503-123503-7 Publisher: The American Physical Society

Bassett, Bruce A.; Tsujikawa, Shinji Inflationary preheating and primordial black holes Journal Physical Review Date: Jun 15 2001 Abstract: Abstract available at publisher web site Journal volume: 63 Issue number: 12 Page number: 123503-123503-7 Publisher: American Physical Society

Basu, D.; Haque-Copilah, S.; Valtonen, M. Blueshifted Quasars Associated with Nearby Galaxies? Journal International Journal of Modern Physics A Date: Mar 20 2000 Abstract: It is possible that supermassive black holes are ejected from centers of galaxies at high speeds. If the ejection happens in a nearby galaxy, then some black holes will travel towards us and may appear quasar-like with a blueshifted spectrum due to the Doppler effect. But quasar spectra are as a rule interpreted as having been redshifted even when there is an equally good or better case for a blueshifted spectrum. Here the authors study the quasars which are apparently associated with galaxies.

Baumgarte, Thomas W. Innermost stable circular orbit of binary black holes Journal Physical Review Date: Jul 15 2000 Abstract: Abstract available at publisher web site Journal volume: 62 Issue number: 2 Page number: 024018-024018-8 Publisher: American Physical Society

Becker, K.; Becker, M. Quantum gravity corrections for Schwarzschild black holes Journal Physical Review, Date: Jul 01 1999 Abstract: The authors consider the matrix theory proposal describing 11-dimensional Schwarzschild black holes. The authors argue that the Newtonian potential between two black holes receives a genuine long-range quantum gravity correction, which is finite and can be computed from the supergravity point of view. In this large-distance limit, the black hole will be treated as a massive pointlike object. The supergravity result agrees with matrix theory up to a numerical factor which the authors have not computed. Journal volume: 60 Issue number: 2 Page number: pp.026003

Behrndt, K.; Cvetic, M.; Sabra, W.A. Entropy of near-extreme N=2 black holes Journal Physical Review, D (Particles Fields) Date: Oct 31 1998 Abstract: The authors give an explicit form of the classical entropy for four-dimensional static near-Bogomol'nyi-Prasad-Sommerfeld-(BPS)-s aturated black holes of N=2 superstring vacua. The expression is obtained by determining the leading corrections in the non-extremality parameter to the corresponding BPS-saturated black hole solutions. Journal volume: 58 Issue number: 8 Page number: 084018 Subjects: Entropy; Superstring Models; Vacuum States; Corrections

Behrndt, K.; Mohaupt, T. Entropy of N=2 black holes and their M-brane description Journal Physical Review, Date: Aug 31 1997 Abstract: In this paper the authors discuss the M-brane description for an N=2 black hole. This solution is a result of the compactification of M-5-brane configurations over a Calabi-Yau threefold with arbitrary intersection numbers C_{ABC}. In analogy with the D-brane description where one counts open string states the authors count here open M-2-branes which end on the M-5-brane. Journal volume: 564 Page number: 2206-2211 Subject keyword: Black Holes; Compactification; Entropy; Solitons; String Models; Supergravity; Composite Models; Extended Particle Model; Field Theories; Mathematical Models; Particle Models; Physical Properties; Quark Model; Quasi Particles; Thermodynamic Properties

Behrndt, K; Mohaupt, T. Entropy of N=2 black holes and their M-brane description Journal Physical Review, D (Particles Fields) Date: Aug 31 1997 Abstract: In this paper the authors discuss the M-brane description for an N=2 black hole. This solution is a result of the compactification of M-5-brane configurations over a Calabi-Yau threefold with arbitrary intersection numbers C_{ABC}. In analogy with the D-brane description where one counts open string states the authors count here open M-2-branes which end on the M-5-brane. Journal volume: 56 Issue number: 4 Page number: 2206-2211 Subjects: Black Holes/Entropy; Entropy; Compactification; String Models; Solitons; Supergravity

Behrndt, Klaus; Bergshoeff, Eric; Halbersma, Rein; Schaar, Jan Pieter.van.der On domain-wall/QFT dualities in various dimensions Journal Classical and Quantum Gravity Date: Nov 01 1999 Abstract: The authors investigate domain-wall/quantum field-theory correspondences in various dimensions. This general analysis covers not only the well studied cases in 10 and 11 dimensions, but also enables us to discuss new cases like a type I/heterotic 6-brane in 10 dimensions and domain-wall dualities in lower than 10 dimensions. The examples the authors discuss include 'd-branes' in six dimensions preserving eight supersymmetries and extreme black holes in various dimensions. In the latter case the authors construct the quantum mechanics Hamiltonian and discuss several limits. Journal volume: 16 Issue number: 11 Page number: 3517-3552 Publisher: Institute of Physics

Bekenstein, Jacob D.; Mayo, Avraham E. Black hole polarization and new entropy bounds Journal Physical Review Date: Jan 15 2000 Abstract: Abstract available at publisher web site Journal volume: 61 Issue number: 2 Page number: 024022-024022-8 Publisher: American Physical Society

Belgiorno, F.; Martellini, M.; Baldicchi, M. Naked Reissner-Nordström singularities and the anomalous magnetic moment of the electron field Journal Physical Review Date: Oct 15 2000 Abstract: Abstract available at publisher web site Journal volume: 62 Issue number: 8 Page number: 084014-084014-11 Publisher: American Physical Society

Beloborodov, Andrei M.; Illarionov, Andrei F. Small-scale inviscid accretion discs around black holes Journal Monthly Notices of the Royal Astronomical Society Date: May 01 2001 Abstract: Gas falling quasi-spherically on to a black hole forms an inner accretion disc if its specific angular momentum l exceeds l ~r g c, where r g is the Schwarzschild radius. The standard disc model assumes ll . The authors argue that, in many black hole sources, accretion flows have angular momenta just above the threshold for disc formation, ll , and assess the accretion mechanism in this regime. In a range l <l<l cr, a small-scale disc forms in which gas spirals fast into the black hole without any help from horizontal viscous stresses.

Berezin, Victor Towards a Theory of Quantum Black Holes Journal International Journal of Modern Physics A Date: Mar 01 2002 Abstract: The authors describe some specific quantum black hole model. It is pointed out that the origin of a black hole entropy is the very process of quantum gravitational collapse. The quantum black hole mass spectrum is extracted from the mass spectrum of the gravitating source. The classical analog of quantum black hole is constructed. Journal volume: 17 Issue number: 06n07 Rights: World Scientific Publishing Company Publisher: World Scientific Publishing Company

Berezin, Victor; Okhrimenko, Maxim A theory of thin shells with orbiting constituents Journal Classical and Quantum Gravity Date: Jun 07 2001 Abstract: P The self-gravitating spherically symmetric thin shells built of orbiting particles are studied. Two new features are found. It is the minimal possible value for the angular momentum of particles, above which elliptic orbits become possible. And the coexistence of both the wormhole solutions and the elliptic or hyperbolic orbits for the same values of parameters (but different initial conditions). Possible applications of this results to astrophysics and quantum black holes are briefly discussed. Journal volume: 18 Issue number: 11 Page number: 2195-2215 Publisher: Institute of Physics

Bernido, C.C.; Aguarte, G. Summation over histories for a particle in spherical orbit around a black hole Journal Physical Review, D (Particles Fields) Date: Aug 31 1997 Abstract: An exact path integral treatment of a relativistic scalar particle in a spherical orbit around a Reissner-Nordström and Schwarzschild black hole is presented. A closed form for the Green function and the energy spectrum are obtained. Journal volume: 56 Issue number: 4 Page number: 2445-2448 Subjects: Black Holes/scalar fields; Scalar Fields/feynman path integral; Orbits; Green Function; Energy Spectra; Propagator; Klein-Gordon Equation; Quantization; Quantum Gravity; Gravitational Radiation; Schwarzschild Metric

Bernido, C.C.; Aguarte, G. Summation over histories for a particle in spherical orbit around a black hole Journal Physical Review, D (Particles Fields) Date: Aug 31 1997 Abstract: An exact path integral treatment of a relativistic scalar particle in a spherical orbit around a Reissner-Nordström and Schwarzschild black hole is presented. A closed form for the Green function and the energy spectrum are obtained. Journal volume: 56 Issue number: 4 Page number: 2445-2448 Subjects: Black Holes/scalar fields; Scalar Fields/feynman path integral; Orbits;

Green Function; Energy Spectra; Propagator; Klein-Gordon Equation; Quantization; Quantum Gravity; Gravitational Radiation; Schwarzschild Metric

Bernido, C.C.; Aguarte, G. Summation over histories for a particle in spherical orbit around a black hole Journal Physical Review, Date: Aug 31 1997 Abstract: An exact path integral treatment of a relativistic scalar particle in a spherical orbit around a Reissner-Nordström and Schwarzschild black hole is presented. A closed form for the Green function and the energy spectrum are obtained. Journal volume: 564 Page number: 2445-2448 Subjects: Black Holes-- Scalar Fields; Scalar Fields-- Feynman Path Integral; Energy Spectra; Gravitational Radiation; Green Function; Klein-Gordon Equation; Orbits; Propagator; Quantization; Quantum Gravity; Schwarzschild Metric; Differential Equations; Equations; Field Theories; Functions; Integrals; Metrics; Partial Differential Equations; Quantum Field Theory; Radiations; Spectra; Wave Equations

Bertolini, Matteo; Trigiante, Mario Regular R-R and NS-NS BPS Black Holes Journal International Journal of Modern Physics A Date: Dec 20 2000 Abstract: The authors show in a precise group theoretical fashion how the generating solution of regular BPS black holes of N=8 supergravity, which is known to be a solution also of a simpler N=2 STU model truncation, can be characterized as purely NS-NS or R-R charged according to the way the corresponding STU model is embedded in the original N=8 theory. Of particular interest is the class of embeddings which yield regular BPS black hole solutions carrying only R-R charge and whose microscopic description can possibly be given in terms of bound states of D-branes only. The microscopic interpretation of the bosonic fields in this class of STU models relies on the solvable Lie algebra (SLA) method. In the present paper, the authors improve this mathematical technique in order to provide two distinct descriptions for type IIA and type IIB theories and an algebraic characterization of S×T-dual embeddings within the N=8, d=4 theory. This analysis willbe applied to the particular example of a four parameter (dilatonic) solution of which both the full macroscopic and microscopic descriptions will be worked out. Journal volume: 15 Issue number: 31 Rights: World Scientific Publishing Company Publisher: World Scientific Publishing Company

Bethe, H.A.; Brown, G.E. Evolution of Binary Compact Objects That Merge Journal Astrophysical Journal Date: Oct 31 1998 Abstract: Beginning from massive binaries in the Galaxy, the authors evolve black hole–neutron star (BH-NS) binaries and binary neutron stars, such as the Hulse-Taylor system PSR 1913+16. The new point in our evolution is a quantitative calculation of the accretion of matter by a neutron star in a common-envelope evolution that sends it into a black hole. The authors calculate the mass of the latter to be {approximately}2.4 M{sub {circle_dot}}. The black hole fate of the first neutron star can only be avoided if the neutron star does not go through common-envelope evolution. Journal volume: 506 Issue number: 2 Page number: 780-789 Subjects: Star Evolution; Star Accretion; Binary Stars; Neutron Stars; Black Holes; Pulsars; Supernovae; Mass Transfer; Gravitational Waves; Gravitation; Statistics

Bethe, H.A.; Brown, G.E. Contribution of High-Mass Black Holes to Mergers of Compact Binaries Journal Astrophysical Journal Date: May 01 1999 Abstract: The authors consider the merging of compact binaries consisting of a high-mass black hole and a neutron star. From stellar evolutionary calculations that include mass loss, the authors estimate that a zero-age main sequence (ZAMS) mass of {approx_gt}80 M{sub {circle_dot}} is necessary before a high-mass black hole can result from a massive O star progenitor. The authors first consider how Cyg X-1, with its measured orbital radius of {approximately} 17 R{sub {circle_dot}}, might evolve. Although this radius is substantially less than the initial distance of two O stars, it is still so large that the resulting compact objects will merge only if an eccentricity close to unity results from a high kick velocity of the neutron star in the final supernova explosion. The authors find that the high-mass black hole neutron star systems contribute substantially to the predicted observational frequency of gravitational waves. The authors discuss how our high-mass black hole formation can be reconciled with the requirements of nucleosynthesis, and the authors indicate that a bimodal distribution of masses of black holes in single stars can account, at least qualitatively, for the many transient sources that contain high-mass black holes. Journal volume: 517 Issue number: 1 Page number: pp.318-327

Bhattacharyya, S.; V., A.; Bombaci, I. Temperature profiles of accretion discs around rapidly rotating strange stars in general relativity: A comparison with neutron stars Journal Astronomy and Astrophysics Date: Apr 12 2001 Abstract: The authors compute the temperature profiles of accretion discs around rapidly rotating strange stars, using constant gravitational mass equilibrium sequences of these objects, considering the full effect of general relativity. Beyond a certain critical value of stellar angular momentum (J), the authors observe the radius ($r_{\rm orb}$) of the innermost stable circular orbit (ISCO) to increase with J (a property seen neither in rotating black holes nor in rotating neutron stars). The reason for this is traced to the crucial dependence of ${\rm d}r_{\rm orb}/{\rm d}J$ on the rate of change of the radial gradient of the Keplerian angular velocity at $r_{\rm orb}$ with respect to J. The structure parameters and temperature profiles obtained are compared with those of neutron stars, as an attempt to provide signatures for distinguishing between the two.

Bicak, J.; Podolsky, J. Global structure of Robinson-Trautman radiative space-times with cosmological constant Journal Physical Review, D (Particles Fields) Date: Feb 28 1997 Abstract: Robinson-Trautman radiative space-times of Petrov type II with a nonvanishing cosmological constant {Lambda} and mass parameter m{gt}0 are studied using analytical methods. They are shown to approach the corresponding spherically symmetric Schwarzschild–de Sitter or Schwarzschild–anti-de Sitter solution at large retarded times. Journal volume: 55 Issue number: 4 Page number: 1985-1993 Subject Keyword: Space-Time/Global Analysis; Cosmology; Space-Time; Cosmological Constant; Mass; Schwarzschild Metric; Gravitational Waves; Black Holes; General Relativity Theory

Billó, Marco; Cacciatori, Sergio; Denef, Frederik; Fré, Pietro; Proeyen, Antoine.Van; Zanon, Daniela The 0-brane action in a general D = 4 supergravity background Journal Classical and Quantum Gravity Date: Jul 01 1999 Abstract: The authors begin by presenting the superparticle action in the background of = 2, D = 4 supergravity coupled to n vector multiplets interacting via an arbitrary special Kähler geometry. Our construction is based on implementing -supersymmetry. In particular, our result can be interpreted as the source term for = 2 BPS black holes with a finite horizon area. When the vector multiplets can be associated with the complex structure moduli of a Calabi-Yau manifold, our 0-brane action can then be derived by wrapping 3-branes around 3-cycles of the 3-fold.

Birmingham, Danny Topological black holes in anti-de Sitter space Journal Classical and Quantum Gravity Date: Apr 01 1999 Abstract: The authors consider a class of black hole solutions to Einstein's equations in d dimensions with a negative cosmological constant. These solutions have the property that the horizon is a (d - 2)-dimensional Einstein manifold of positive, zero or negative curvature. The mass, temperature and entropy are calculated. Using the correspondence with conformal field theory, the phase structure of the solutions is examined, and used to determine the correct mass dependence of the Bekenstein-Hawking entropy. Journal volume: 16 Issue number: 4 Page number: 1197-1205 Publisher: Institute of Physics

Birmingham, Danny; Sen, Siddhartha Exact black hole entropy bound in conformal field theory Journal Physical Review Date: Feb 15 2001 Abstract: Abstract available at publisher web site Journal volume: 63 Issue number: 4 Page number: 047501-047501-3 Publisher: American Physical Society

Birmingham, Danny; Sen, Siddhartha Gott Time Machines, BTZ Black Hole Formation, and Choptuik Scaling Journal Physical Review Letters Date: Feb 07 2000 Abstract: Abstract available at publisher web site Journal volume: 84 Issue number: 6 Page number: 1074-1077 Publisher: American Physical Society

Bizon', Piotr; Chmaj, Tadeusz Remark on the formation of colored black holes via fine-tuning Journal Physical Review Date: Mar 15 2000 Abstract: Abstract available at publisher web site Journal volume: 61 Issue number: 6 Page number: 067501-067501-2 Publisher: American Physical Society

Bjoraker, Jeff; Hosotani, Yutaka Monopoles, dyons, and black holes in the four-dimensional Einstein-Yang-Mills theory Journal Physical Review. D, Particles Fields Date: Aug 15 2000 Abstract: A continuum of monopole, dyon, and black hole solutions exists in the Einstein-Yang-Mills theory in asymptotically anti-de Sitter space. Their structure is studied in detail. The solutions are classified by non-Abelian electric and magnetic charges and the Arnowitt-Deser-Misner mass. The stability of the solutions which have no node in non-Abelian magnetic fields is established.

Bjoraker, Jeff; Hosotani, Yutaka Monopoles, dyons, and black holes in the four-dimensional Einstein-Yang-Mills theory Journal Physical Review Date: Aug 15 2000 Abstract: Abstract available at publisher web site Journal volume: 62 Issue number: 4 Page number: 043513-043513-15 Publisher: American Physical Society

Bojowald, M.; Kastrup, H. A.; Schramm, F.; Strobl, T. Group theoretical quantization of a phase space $S1 \times R+$ and the mass spectrum of Schwarzschild black holes in D space-time dimensions Journal Physical Review Date: Aug 15 2000 Abstract: Abstract available at publisher web site Journal volume: 62 Issue number: 4 Page number: 044026-044026-20 Publisher: American Physical Society

Boldt, Elihu; Ghosh, Pranab Cosmic rays from remnants of quasars? Journal Monthly Notices of the Royal Astronomical Society Date: Aug 01 1999 Abstract: Considerations of the collision losses for protons traversing the 2.7-K blackbody microwave radiation field have led to the conclusion that the highest energy cosmic rays, those observed at 1020 eV, must come from sources within the present epoch. In light of this constraint, it is here suggested that these particles may be accelerated near the event horizons of spinning supermassive black holes associated with presently inactive quasar remnants. The required emf is generated by the black hole induced rotation of externally supplied magnetic field lines threading the horizon. Producing the observed flux of the highest energy cosmic rays would constitute a negligible drain on the black hole dynamo.

Bonnor, W.B. A model of a spheroidal body Journal Classical and Quantum Gravity Date: Feb 01 1998 Abstract: Electrically counterpoised dust (ECD) is incoherent electrically charged matter in which the charge density is equal to the mass density in relativistic units. Einstein's equations admit static solutions modelling bodies of arbitrary shape composed of ECD. In this paper I present a model of a non-singular prolate spheroid of ECD, and consider its size and mass in relation to the hoop conjecture and to the isoperimetric conjecture for black holes. The conclusion is that it satisfies the hoop conjecture, but, for any given mass, the surface area can be arbitrarily small. Journal volume: 15 Issue number: 2 Page number: 351-356 Publisher: Institute of Physics

Booth, I. S.; Mann, R. B. Static and infalling quasilocal energy of charged and naked black holes Journal Physical Review Date: Dec 15 1999 Abstract: Abstract available at publisher web site Journal volume: 60 Issue number: 12 Page number: 124009-124009-22 Publisher: American Physical Society

Borde, A. Regular black holes and topology change Journal Physical Review, D (Particles Fields) Date: Jun 30 1997 Abstract: The conditions are clarified under which regular (i.e., singularity-free) black holes can exist. It is shown that in a large class of spacetimes that satisfy the weak energy condition the existence of a regular black hole requires topology change. Journal volume: 55 Issue number: 12 Page number: 7615-7617 Subjects: Black Holes/topology; Topology; Singularity; Space-Time; Energy

Borde, A. Regular black holes and topology change Journal Physical Review, D (Particles Fields) Date: Jun 30 1997 Abstract: The conditions are clarified under which regular (i.e., singularity-free) black holes can exist. It is shown that in a large class of spacetimes that satisfy the weak energy condition the existence of a regular black hole requires topology change. Journal volume: 55 Issue number: 12 Page number: 7615-7617 Subjects: Black Holes/topology; Topology; Singularity; Space-Time; Energy

Borde, A. Regular black holes and topology change Journal Physical Review, Date: Jun 30 1997 Abstract: The conditions are clarified under which regular (i.e., singularity-free) black holes can exist. It is shown that in a large class of spacetimes that satisfy the weak energy condition the existence of a regular black hole requires topology change. Journal volume: 5512 Page number: 7615-7617

Bose, S.; Parker, L.; Peleg, Y. Lorentzian approach to black hole thermodynamics in the Hamiltonian formulation Journal Physical Review, Date: Jul 31 1997 Abstract: In this work, the authors extend the analysis of Brown and York to find the quasilocal energy in a spherical box in the Schwarzschild black-hole spacetime. In such a case, quasilocal energy is the value of the Hamiltonian that generates unit magnitude proper-time translations on the box orthogonal to the spatial hypersurfaces foliating the Schwarzschild spacetime. The authors call this Hamiltonian the Brown-York Hamiltonian. Journal volume: 562 Page number: 987-1000 Subjects: Black Holes-- Thermodynamics; Thermodynamics-- Hamiltonians; Boundary Conditions; Energy; Free Energy; Geometry; Partition Functions; Schwarzschild Metric; Space-Time; Energy; Functions; Mathematical Operators; Mathematics; Metrics; Physical Properties; Quantum Operators; Thermodynamic Properties

Bose, S.; Parker, L; Peleg, Y. Lorentzian approach to black hole thermodynamics in the Hamiltonian formulation Journal Physical Review, D (Particles Fields) Date: Jul 31 1997 Abstract: In this work, the authors extend the analysis of Brown and York to find the quasilocal energy in a spherical box in the Schwarzschild black-hole spacetime. In such a case, quasilocal energy is the value of the Hamiltonian that generates unit magnitude proper-time translations on the box orthogonal to the spatial hypersurfaces foliating the Schwarzschild spacetime. The authors call this Hamiltonian the Brown-York Hamiltonian. The authors find different classes of foliations that correspond to time evolution by the Brown-York Hamiltonian. Journal volume: 56 Issue number: 2 Page number: 987-1000 Subjects: Black Holes/thermodynamics; Thermodynamics/hamiltonians; Schwarzschild Metric; Geometry

Bose, Sukanta; Naing, Thant Zin Quasilocal energy for rotating charged black hole solutions in general relativity and string theory Journal Physical Review Date: Nov 15 1999 Abstract: The authors first consider the charged Kerr black hole. For such a spacetime, the authors calculate the quasilocal energy within a two-surface of constant Boyer-Lindquist radius embedded in a constant stationary-time slice. Keeping with Martinez's conjecture, at the outer horizon this energy equals radicand: A /radical: /2 radicand: . The energy is positive and monotonically decreases to the ADM mass as the boundary-surface radius diverges. Next the authors perform an analogous calculation for the quasilocal energy for the Kerr-Sen spacetime, which corresponds to four-dimensional rotating charged black hole solutions in heterotic string theory. The behavior of this energy as a function of the boundary-surface radius is similar to the charged Kerr case. However, the authors show that it does not approach the expression conjectured by Martinez at the horizon. Journal volume: 60 Issue number: 10 Page number: 104027-104027-9 Publisher: American Physical Society

Bousso, R. Pair creation of dilaton black holes in extended inflation Journal Physical Review, D (Particles Fields) Date: Apr 30 1997 Abstract: Dilatonic charged Nariai instantons mediate the nucleation of black hole pairs during extended chaotic inflation. Depending on the dilaton and inflaton fields, the black holes are described by one of two approximations in the Lorentzian regime. For each case the authors find Euclidean solutions that satisfy the no boundary proposal. Journal volume: 55 Issue number: 8 Page number: 4889-4897 Subjects: Black Holes/pair production; Cosmology; Inflation; Nucleation; Cosmological Constant; Inflationary Universe

Bousso, Raphael The holographic principle for general backgrounds Journal Classical and Quantum Gravity Date: Mar 07 2000 Abstract: The authors aim to establish the holographic principle as a universal law, rather than a property only of static systems and special spacetimes. Our covariant formalism yields an upper bound on entropy which applies to both open and closed surfaces, independently of shape or location. It reduces to the Bekenstein bound whenever the latter is expected to hold, but complements it with novel bounds when gravity dominates. In particular, it remains valid in closed FRW cosmologies and in the interior of black holes.

Brandt, C. F. C.; Lin, L.-M.; da Rocha, J. F. Villas; Wang, A. Z.; Fang, L. Z. Gravitational Collapse of Spherically Symmetric Perfect Fluidwith Kinematic Self-Similarity Journal International Journal of Modern Physics A Date: Feb 01 2002 Abstract: Analytic spherically symmetric solutions of the Einstein fieldequations coupled with a perfect fluid and with self-similarities ofthe zeroth, first and second kinds, found recently by Benoit and Coley[Class. Quantum Grav.15, 2397 (1998)], are studied,and found that some of them represent gravitational collapse. When thesolutions have self-similarity of the first (homothetic) kind, some ofthe solutions may represent critical collapse but in the sense thatnow the "critical" solution separates the collapse that forms blackholes from the collapse that forms naked singularities.

Brandt, Steve; Correll, Randall; Gómez, Roberto; Huq, Mijan; Laguna, Pablo; Lehner, Luis; Marronetti, Pedro; Matzner, Richard A.; Neilsen, David; Pullin, Jorge; Schnetter, Erik; Shoemaker, Deirdre; Winicour, Jeffrey Grazing Collisions of Black Holes via the Excision of Singularities Journal Physical Review Letters Date: Dec 25 2000 Abstract: Abstract available at publisher web site Journal volume: 85 Issue number: 26 Page number: 5496-5499 Publisher: American Physical Society

Brandt, Steven; Anninos, Peter Radiation recoil from highly distorted black holes Journal Physical Review Date: Oct 15 1999 Abstract: Abstract available at publisher web site Journal volume: 60 Issue number: 8 Page number: 084005-084005-8 Publisher: American Physical Society

Brax, Philippe; Mandal, Gautam; Oz, Yaron Supergravity description of non-Bogomol'nyi-Prasad-Sommerfield branes Journal Physical Review Date: Mar 15 2001 Abstract: Abstract available at publisher web site overline: Dp -branes. The authors study the physical properties of the solutions and analyze the supergravity description of tachyon condensation. The authors construct an interpolation between the brane-antibrane solution and the Schwarzschild solution and discuss its possible application to the study of non-supersymmetric black holes. Journal volume: 63 Issue number: 6 Page number: 064008-064008-14 Publisher: American Physical Society

Breckenridge, J.C; Michaud, G; Myers, R.C. More D-brane bound states Journal Physical Review, D (Particles Fields) Date: May 31 1997 Abstract: The low-energy background field solutions corresponding to D-brane bound states which possess a difference in dimension of two are presented. These solutions are constructed using the T-duality map between the type IIA and IIB superstring theories. Journal volume: 55 Issue number: 10 Page number: 6438-6446 Subjects: Superstring Models/Bound Dtate; Cosmology; Duality; Supersymmetry; Solitons; Cosmology; Black Holes; Dimensions

Breckenridge, J.C; Michaud, G; Myers, R.C. New angles on D-branes Journal Physical Review, D (Particles Fields) Date: Oct 31 1997 Abstract: A low-energy background field solution is presented which describes several D-membranes oriented at angles with respect to one another. The mass and charge densities for this configuration are computed and found to saturate the Bogomol'nyi-Prasad-Sommerfield bound, implying the preservation of one-quarter of the supersymmetries. T duality is exploited to construct new solutions with nontrivial angles from the basic one. Journal volume: 56 Issue number: 8 Page number: 5172-5178 Subjects: Solitons/string models; T Invariance; Supersymmetry; Solitons; Mass; Charge Density; Duality; Black Holes; Entropy

Brevik, I.; Halnes, G. Light rays at optical black holes in moving media Journal Physical Review Date: Jan 15 2002 Abstract: Abstract available at publisher web site Journal volume: 65 Issue number: 2 Page number: 024005-024005-12 Publisher: American Physical Society

Brihaye, Y.; Hartmann, B.; Kunz, J. Dilatonic monopoles and "hairy" black holes Journal Physical Review Date: Jan 15 2002 Abstract: Abstract available at publisher web site Journal volume: 65 Issue number: 2 Page number: 024019-024019-10 Publisher: American Physical Society

Brihaye, Yves; Hartmann, Betti; Kunz, Jutta Gravitating monopoles and black holes at intermediate Higgs boson masses Journal Physical Review Date: Aug 15 2000 Abstract: Abstract available at publisher web site Journal volume: 62 Issue number: 4 Page number: 044008-044008-6 Publisher: American Physical Society

Brihaye, Yves; Hartmann, Betti; Kunz, Jutta; Tell, Nade`ge Dyonic non-Abelian black holes Journal Physical Review Date: Nov 15 1999 Abstract: Abstract available at publisher web site Journal volume: 60 Issue number: 10 Page number: 104016-104016-13 Publisher: American Physical Society

Bringmann, Torsten; Kiefer, Claus; Polarski, David Primordial black holes from inflationary models with and without broken scale invariance Journal Physical Review Date: Jan 15 2002 Abstract: Abstract available at publisher web site Journal volume: 65 Issue number: 2 Page number: 024008-024008-7 Publisher: American Physical Society

Bromley, B.C.; Miller, W.A.; Pariev, V.I.; Pariev, V.I. Bounds on the inner radius of emission around supermassive black holes Journal AIP Conference Proceedings Date:

Apr 30 1998 Abstract: Observations of iron K-alpha fluorescence lines in Seyfert galaxies provide strong evidence for an accretion disk around a supermassive black hole as the source of the line emission. Here the authors consider a diagnostic of a disk line profile, based on the minimum and maximum frequency of emission, to robustly constrain the inner radius of the disk. This diagnostic is applied to spectra from Seyfert I galaxies (Iwasawa *et al.* 1996 [1]; Nandra *et al.* 1997 [2]), and a composite Seyfert II spectrum (Turner *et al.* 1997 [3]). Journal volume: 431 Issue number: 1 Page number: 269-272 Subjects: X-Ray Spectra; Fluorescence; Line Widths; Seyfert Galaxies; Accretion Disks

Bronnikov, K. A. Regular magnetic black holes and monopoles from nonlinear electrodynamics Journal Physical Review Date: Feb 15 2001 Abstract: Abstract available at publisher web site Journal volume: 63 Issue number: 4 Page number: 044005-044005-6 Publisher: American Physical Society

Bronnikov, K. A. Spherically symmetric false vacuum: No-go theorems and global structure Journal Physical Review Date: Sep 15 2001 Abstract: Abstract available at publisher web site Journal volume: 64 Issue number: 6 Page number: 064013-064013-4American Physical Society

Brown, J.D. Duality invariance of black hole creation rates Journal Physical Review, D (Particles Fields) Date: Jul 31 1997 Abstract: Pair creation of electrically charged black holes and its dual process, pair creation of magnetically charged black holes, are considered. It is shown that the creation rates are equal provided the boundary conditions for the two processes are dual to one another. This conclusion follows from a careful analysis of boundary terms and boundary conditions for the Maxwell action. Journal volume: 56 Issue number: 2 Page number: 1001-1004 Subjects: Duality; Charges; Boundary Conditions; Einstein-Maxwell Equations; Quantum Gravity

Bullock, J.S.; Primack, J.R. Non-Gaussian fluctuations and primordial black holes from inflation Journal Physical Review, Date: Jun 30 1997 Abstract: The authors explore the role of non-Gaussian fluctuations in primordial black hole (PBH) formation and show that the standard Gaussian assumption, used in all PBH formation papers to date, is not justified. Since large spikes in power are usually associated with flat regions of the inflaton potential, quantum fluctuations become more important in the field dynamics, leading to mode-mode coupling and non-Gaussian statistics. Journal volume: 5512 Page number: 7423-7439 Subjects: Black Holes-- Fluctuations; Cosmology; Distribution; Inflation; Numerical Analysis; Probability; Simulation; Stochastic Processes; Mathematics; Variations

Bullock, J.S; Primack, J.R. Non-Gaussian fluctuations and primordial black holes from inflation Journal Physical Review, D (Particles Fields) Date: Jun 30 1997 Abstract: The authors explore the role of non-Gaussian fluctuations in primordial black hole (PBH) formation and show that the standard Gaussian assumption, used in all PBH formation papers to date, is not justified. Since large spikes in power are usually associated with flat regions of the inflaton potential, quantum fluctuations become more important in the field dynamics, leading to mode-mode coupling and non-Gaussian statistics. Journal volume: 55 Issue number: 12 Page number: 7423-7439 Subjects: Cosmology; Stochastic Processes; Numerical Analysis; Fluctuations; Inflation; Distribution; Simulation; Probability

Buonanno, Alessandra; Damour, Thibault Transition from inspiral to plunge in binary black hole coalescences Journal Physical Review Date: Sep 15 2000 Abstract: Abstract available at publisher web site Journal volume: 62 Issue number: 6 Page number: 064015-064015-24 Publisher: American Physical Society

Burko, L.M. QED blue-sheet effects inside black holes Journal Physical Review, D (Particles Fields) Date: Feb 28 1997 Abstract: The interaction of the unboundedly blueshifted photons of the cosmic microwave background radiation with a physical object falling towards the inner horizon of a Reissner-Nordström black hole is analyzed. To evaluate this interaction the authors consider the QED effects up to the second order in the perturbation expansion. Journal volume: 55 Issue number: 4 Page number: 2105-2109 Subjects: Perturbation Theory; Background Radiation; Microwave Radiation; Photons; Space-Time

Burko, L.M. Structure of the Black Hole[close quote]s Cauchy-Horizon Singularity Journal Physical Review Letters Date: Dec 31 1997 Abstract: The authors study the Cauchy-horizon (CH) singularity of a spherical charged black hole perturbed nonlinearly by a self-gravitating massless scalar field. The authors show numerically that the singularity is weak both at the early and at the late sections of the CH, where the focusing of the area coordinate r is strong. In the early section the metric is almost Reissner-Nordström, and the fields behave according to perturbation analysis. Journal volume: 7925 Page number: 4958-4961 Subject keyword: Black Holes; Einstein Field Equations; General Relativity Theory; Gravitation; Massless Particles; Numerical Analysis

Burko, L.M. Structure of the Black Hole's Cauchy-Horizon Singularity Journal Physical Review Letters Date: Dec 31 1997 Abstract: The authors study the Cauchy-horizon (CH) singularity of a spherical charged black hole perturbed nonlinearly by a self-gravitating massless scalar field. The authors show numerically that the singularity is weak both at the early and at the late sections of the CH, where the focusing of the area coordinate r is strong. In the early section the metric is almost Reissner-Nordström, and the fields behave according to perturbation analysis. Journal volume: 79 Issue number: 25 Page number: 4958-4961 Subject Keyword: Black Holes/Singularity; Singularity; Scalar Fields; Massless Particles; Numerical Analysis; Gravitation; General Relativity Theory; Einstein Field Equations; Perturbation Theory

Burko, Lior M. Strength of the null singularity inside black holes Journal Physical Review Date: Nov 15 1999 Abstract: Abstract available at publisher web site Journal volume: 60 Issue number: 10 Page number: 104033-104033-5 Publisher: American Physical Society

Cadeau, C.; Woolgar, E. New five-dimensional black holes classified by horizon geometry, and a Bianchi VI braneworld Journal Classical and Quantum Gravity Date: Feb 07 2001 Abstract: P The authors introduce two new families of solutions to the vacuum Einstein equations with negative cosmological constant in five dimensions. These solutions are static black holes whose horizons are modelled on the 3-geometries nilgeometry and solvegeometry. Thus the horizons (and the exterior spacetimes) can be foliated by compact 3-manifolds that are neither spherical, toroidal, hyperbolic, nor product manifolds, and therefore are of a topological type not previously encountered in black hole solutions. As an application, the authors use the solvegeometry solutions to construct Bianchi VI -1 braneworld cosmologies. Journal volume: 18 Issue number: 3 Page number: 527-542 Publisher: Institute of Physics

Cadoni, M.; Carta, P.; Cavaglia & Grave;, M.; Mignemi, S. Conformal dynamics of 0-branes Journal Physical Review Date: Jan 15 2002 Abstract: Abstract available at publisher web site Journal volume: 65 Issue number: 2 Page number: 024002-024002-12 Publisher: American Physical Society

Cadoni, M.; Carta, P.; Klemm, D.; Mignemi, S. 2D anti-de Sitter gravity as a conformally invariant mechanical system Journal Physical Review Date: Jun 15 2001 Abstract: Abstract available at publisher web site Journal volume: 63 Issue number: 12 Page number: 125021-125021-5 Publisher: American Physical Society

Cadoni, M.; Carta, P.; Klemm, D.; Mignemi, S. 2D anti–de Sitter gravity as a conformally invariant mechanical system Journal Physical Review Date: Jun 15 2001 Abstract: The authors show that two-dimensional (2D) AdS gravity induces on the spacetime boundary a conformally invariant dynamics that can be described in terms of a de Alfaro–Fubini–Furlan model coupled to an external source with conformal dimension 2. The external source encodes information about the gauge symmetries of the 2D gravity system. Alternatively, there exists a description in terms of a mechanical system with anholonomic constraints. The considered systems are invariant under the action of the conformal group generated by a Virasoro algebra, which occurs also as an asymptotic symmetry algebra of two-dimensional anti–de Sitter space.

Cadoni, Mariano Dimensional reduction of 4D heterotic string black holes Journal Physical Review Date: Oct 15 1999 Abstract: Abstract available at publisher web site Journal volume: 60 Issue number: 8 Page number: 084016-084016-7 Publisher: American Physical Society

Cadoni, Mariano; Cavaglia, Marco Open strings, 2D gravity, and AdS/CFT correspondence Journal Physical Review Date: Apr 15 2001 Abstract: The authors present a detailed discussion of the duality between dilaton gravity on AdS_2 and open strings. The correspondence between the two theories is established using their symmetries and field theoretical, thermodynamic, and statistical arguments. The authors use the dual conformal field theory to describe two-dimensional black holes. In particular, all the semiclassical features of the black holes, including the entropy, have a natural interpretation in terms of the dual microscopic conformal dynamics. The previous results are discussed in the general framework of the anti--de Sitter/conformal field theory dualities. Journal volume: 63 Issue number: 8 Page number: 084024-084024-12 Publisher: The American Physical Society

Cadoni, Mariano; Cavaglia & Grave;, Marco Open strings, 2D gravity, and AdS/CFT correspondence Journal Physical Review Date: Apr 15 2001 Abstract: Abstract available at publisher web site Journal volume: 63 Issue number: 8 Page number: 084024-084024-12 Publisher: American Physical Society

Cadoni, Mariano; Mignemi, Salvatore Reply to "Comment on 'Entropy of 2D black holes from counting microstates' " Journal Physical Review Date: Apr 15 2000 Abstract: Abstract available at publisher web site Journal volume: 61 Issue number: 8 Page number: 088502-088502-2 Publisher: American Physical Society

Cai, Mingliang; Galloway, Gregory J. On the topology and area of higher-dimensional black holes Journal Classical and Quantum Gravity Date: Jul 21 2001 Abstract: P Over the past decade there has been an increasing interest in the study of black holes, and related objects, in higher (and lower) dimensions, motivated to a large extent by developments in string theory. The aim of the present paper is to obtain higher-dimensional analogues of some well known results for black holes in 3 + 1 dimensions. More precisely, the authors obtain extensions to higher dimensions of Hawking's black hole topology theorem for asymptotically flat ($\Lambda = 0$) black hole spacetimes, and Gibbons' and Woolgar's genus-dependent, lower entropy bound for topological black holes in asymptotically locally anti-de Sitter ($\Lambda<0$) spacetimes. Journal volume: 18 Issue number: 14 Page number: 2707-2718 Publisher: Institute of Physics

Cai, R. Effective Spatial Dimension of Extremal Nondilatonic Black p-Branes and the Description on the World Volume Journal Physical Review Letters Date: Mar 31 1997 Abstract: By investigating the critical behavior appearing at the extremal limit of the nondilatonic, black p-branes in (d+p) dimensions, the authors find that some critical exponents related to the critical point obey the scaling laws. From the scaling laws they obtain that the effective spatial dimension of the nondilatonic black holes and black strings is one, and is p for the nondilatonic black p-branes. Journal volume: 78 Issue number: 13 Page number: 2531-2534 Subjects: Scaling Laws; Space-Time

Cai, Rong-Gen Cardy-Verlinde formula and AdS black holes Journal Physical Review Date: Jun 15 2001 Abstract: In a recent paper by E. Verlinde, hep-th/0008140, an interesting formula has been put forward, which relates the entropy of a conformal formal field in arbitrary dimensions to its total energy and Casimir energy. This formula has been shown to hold for the conformal field theories that have anti–de Sitter (AdS) duals in the cases of AdS Schwarzschild black holes and AdS Kerr black holes. In this paper the authors further check this formula with various black holes with AdS asymptotics..

Cai, Rong-Gen Cardy-Verlinde formula and AdS black holes Journal Physical Review Date: Jun 15 2001 Abstract: Abstract available at publisher web site Journal volume: 63 Issue number: 12 Page number: 124018-124018-9 Publisher: American Physical Society

Cai, Rong-Gen; Ji, Jeong-Young; Soh, Kwang-Sup Action and entropy of black holes in spacetimes with a cosmological constant Journal Classical and Quantum Gravity Date: Sep 01 1998 Abstract: In the Euclidean path-integral approach, the authors calculate the actions and the entropies for the Reissner-Nordström-de Sitter solutions. When the temperatures of black hole and cosmological horizons are equal, the entropy is the sum of one-quarter areas of black hole and cosmological horizons; when the inner and outer black hole horizons coincide, the entropy is only one-quarter area of the cosmological horizon and the entropy vanishes when the two black hole horizons and cosmological horizon coincide. The authors also calculate the Euler numbers of the corresponding Euclidean manifolds, and discuss the relationship between the entropy of instanton and the Euler number. Journal volume: 15 Issue number: 9 Page number: 2783-2793 Publisher: Institute of Physics

Cai, Rong-Gen; Myung, Yun Soo; Ohta, Nobuyoshi Bekenstein bound, holography and brane cosmology in charged black hole backgrounds Journal Classical and Quantum Gravity Date: Dec 21 2001 Abstract: The authors obtain a Bekenstein entropy bound for charged objects in arbitrary dimensions (D 4) using the D-bound recently proposed by Bousso. With the help of thermodynamics of conformal field theories corresponding to anti-de sitter (AdS) Reissner-Norström (RN) black holes, they discuss the relation between the Bekenstein and Bekenstein-Verlinde bounds. In particular, the authors propose a Bekenstein-Verlinde-like bound for the charged systems.

Cai, Rong-Gen; Ohta, Nobuyoshi Surface counterterms and boundary stress-energy tensors for asymptotically non-anti-de Sitter spaces Journal Physical Review Date: Jul 15 2000 Abstract: Abstract available at publisher web site Journal volume: 62 Issue number: 2 Page number: 024006-024006-12 Publisher: American Physical Society

Cai, Rong-Gen; Zhang, Yuan-Zhong Holography and brane cosmology in domain wall backgrounds Journal Physical Review Date: Nov 15 2001 Abstract: Abstract available at publisher web site Journal volume: 64 Issue number: 10 Page number: 104015-104015-8 Publisher: American Physical Society

Caldarelli, Marco M.; Cognola, Guido; Klemm, .Dietmar Thermodynamics of Kerr-Newman-AdS black holes and conformal field theories Journal Classical and Quantum Gravity Date: Jan 21 2000 Abstract: The authors study the thermodynamics of four-dimensional Kerr-

Newman-AdS black holes both in the canonical and the grand-canonical ensemble. The stability conditions are investigated, and the complete phase diagrams are obtained, which include the Hawking-Page phase transition in the grand-canonical ensemble. In the canonical case, one has a first-order transition between small and large black holes, which disappears for a sufficiently large electric charge or angular momentum. This disappearance corresponds to a critical point in the phase diagram.

Cardoso, Gabriel Lopes; Wit, Bernard de; Mohaupt, .Thomas Area law corrections from state counting and supergravity Journal Classical and Quantum Gravity Date: Mar 07 2000 Abstract: Modifications of the Bekenstein-Hawking area law for black holes are crucial in order to find agreement between the microscopic entropy based on state counting and the macroscopic entropy based on an effective field theory computation. The authors discuss this and related issues for the case of four-dimensional $N = 2$ supersymmetric black holes. The authors also briefly comment on the state counting for $N = 4$ and 8 black holes. Journal volume: 17 Issue number: 5 Page number: 1007-1015 Publisher: Institute of Physics

Cardoso, Vitor; Lemos, José P. S. Quasinormal modes of Schwarzschild-anti-de Sitter black holes: Electromagnetic and gravitational perturbations Journal Physical Review Date: Oct 15 2001 Abstract: Abstract available at publisher web site Journal volume: 64 Issue number: 8 Page number: 084017-084017-8 Publisher: American Physical Society

Cardoso, Vitor; Lemos, José P. S. Scalar, electromagnetic, and Weyl perturbations of BTZ black holes: Quasinormal modes Journal Physical Review Date: Jun 15 2001 Abstract: Abstract available at publisher web site Journal volume: 63 Issue number: 12 Page number: 124015-124015-6 Publisher: American Physical Society

Cardoso, Vitor; Lemos, José.P.S. Quasi-normal modes of toroidal, cylindrical and planar black holes in anti-de Sitter spacetimes: scalar, electromagnetic and gravitational perturbations Journal Classical and Quantum Gravity Date: Dec 07 2001 Abstract: The authors study the quasi-normal modes (QNM) of scalar, electromagnetic and gravitational perturbations of black holes in general relativity whose horizons have toroidal, cylindrical or planar topology in an asymptotically anti-de Sitter (AdS) spacetime. The associated QNM frequencies describe the decay in time of the corresponding test field in the vicinities of the black hole.

Cardoso, Vitor; Lemos, José P. S. Scalar, electromagnetic, and Weyl perturbations of BTZ black holes: Quasinormal modes Journal Physical Review Date: Jun 15 2001 Abstract: The authors calculate the quasinormal modes and associated frequencies of the Banados-Teitelboim-Zanelli (BTZ) nonrotating black hole. This black hole lives in 2+1 dimensions in an asymptotically anti–de Sitter spacetime. The authors obtain exact results for the wave function and quasinormal frequencies of scalar, electromagnetic and Weyl (neutrino) perturbations. Journal volume: 63 Issue number: 12 Page number: 124015-124015-6 Publisher: The American Physical Society

Carlip, S. Black Hole Entropy from Conformal Field Theory in Any Dimension Journal Physical Review Letters Date: Apr 30 1999 Abstract: Restricted to a black hole horizon, the "gauge" algebra of surface deformations in general relativity contains a Virasoro subalgebra with a calculable central charge. The fields in any quantum theory of gravity must transform accordingly, i.e., they must admit a conformal field theory description. Journal volume: 82 Issue number: 14 Page number: 2828-2831 Subjects: General Relativity Theory

Carlip, S. Entropy from conformal field theory at Killing horizons Journal Classical and

Quantum Gravity Date: Oct 01 1999 Abstract: On a manifold with a boundary, the constraint algebra of general relativity may acquire a central extension, which can be computed using covariant phase space techniques. When the boundary is a (local) Killing horizon, a natural set of boundary conditions leads to a Virasoro subalgebra with a calculable central charge. Conformal field theory methods may then be used to determine the density of states at the boundary.

Casadio, R; Harms, B. Microfield dynamics of black holes Journal Physical Review, D (Particles Fields) Date: Aug 31 1998 Abstract: The microcanonical treatment of black holes as opposed to the canonical formulation is reviewed and some major differences are displayed. The authors propose a microcanonical alternative to the thermodynamical expression for the number density and discuss its characteristics. Journal volume: 58 Issue number: 4 Page number: 044014 Subjects: Quantum Mechanics; Thermodynamics;

Casadio, R; Harms, B; Leblanc, Y.; Cox, P.H. Perturbations in the Kerr-Newman dilatonic black hole background: Maxwell waves Journal Physical Review, D (Particles Fields) Date: Oct 31 1997 Abstract: In this paper the authors analyze the perturbations of the Kerr-Newman dilatonic black hole background. For this purpose the authors perform a double expansion in both the background electric charge and the wave parameters of the relevant quantities in the Newman-Penrose formalism. The authors then display the gravitational, dilatonic, and electromagnetic equations, which reproduce the static solution (at zero order in the wave parameter) and the corresponding wave equations in the Kerr background (at first order in the wave parameter and zero order in the electric charge). Journal volume: 56 Issue number: 8 Page number: 4948-4961 Subjects: Quantum Gravity; Electric Charges; Maxwell Equations; Corrections; Wave Propagation

Casadio, Roberto; Harms, Benjamin Black hole evaporation and compact extra dimensions Journal Physical Review Date: Jul 15 2001 Abstract: The authors study the evaporation of black holes in space-times with extra dimensions of size L by employing the microcanonical picture of Hawking's radiation. The authors show that the luminosity is greatly damped when the horizon becomes smaller than L and black holes born with an initial size smaller than L are almost stable. This effect is due to the strong dependence of both the occupation number density of Hawking quanta and the greybody factor of a black hole on the dimensionality of space. Although the picture of what happens when the horizon shrinks to a size L is still incomplete, the authors argue that there might occur an outburst of energy which leaves a quasistable remnant. Journal volume: 64 Issue number: 2 Page number: 024016-024016-8 Publisher: The American Physical Society

Casadio, Roberto; Harms, Benjamin Black hole evaporation and compact extra dimensions Journal Physical Review Date: Jul 15 2001 Abstract: Abstract available at publisher web site Journal volume: 64 Issue number: 2 Page number: 024016-024016-8 Publisher: American Physical Society

Castiñeiras, J.; Matsas, G. E. A. Low-energy sector quantization of a massless scalar field outside a Reissner-Nordström black hole and static sources Journal Physical Review Date: Sep 15 2000 Abstract: Abstract available at publisher web site Journal volume: 62 Issue number: 6 Page number: 064001-064001-6 Publisher: American Physical Society

Cataldo, Mauricio; García, Alberto Regular (2+1)-dimensional black holes within nonlinear electrodynamics Journal Physical Review Date: Apr 15 2000 Abstract: Abstract available at publisher web site Journal volume: 61 Issue number: 8 Page number: 084003-084003-5 Publisher: American Physical Society

Cattaneo, Andrea Quasars and galaxy formation Journal Monthly Notices of the Royal Astronomical Society Date: Jun 01 2001 Abstract: Quasars are widely believed to be powered by accretion on to supermassive black holes and there is now considerable evidence for a link between mergers, the activity of quasars and the formation of spheroids. Cattaneo, Haehnelt & Rees have demonstrated that a very simple model in which supermassive black holes form and accrete most of their mass in mergers of galaxies of comparable masses can reproduce the observed relation of black hole mass to bulge luminosity if black holes accrete a fraction of the mass in the merging remnant that varies with redshift. Here the authors investigate whether this simple model can account for the luminosity function of quasars and for the redshift evolution of the quasar population.

Celotti, A.; Miller, J.C.; Sciama, D.W. Astrophysical evidence for the existence of black holes Journal Classical and Quantum Gravity Date: Dec 01 1999 Abstract: Following a short account of the history of the idea of black holes, the authors present a review of the current status of the search for observational evidence of their existence aimed at an audience of relativists rather than astronomers or astrophysicists. The authors focus on two different regimes: that of stellar-mass black holes and that of black holes with the masses of galactic nuclei. Journal volume: 16 Issue number: 12A Page number: A3-A21 Publisher: Institute of Physics

Chakrabarti, Sandip K.; Das, Santabrata Model dependence of transonic properties of accretion flows around black holes Journal Monthly Notices of the Royal Astronomical Society Date: Nov 01 2001 Abstract: The authors analytically study how the behaviour of accretion flows changes when the flow model is varied. The authors study the transonic properties of the conical flow, a flow of constant height and a flow in vertical equilibrium, and show that all these models are basically identical, provided that the polytropic constant is suitably changed from one model to another.

Chakrabarti, Sandip K.; Mukhopadhyay, Banibrata Scattering of Dirac waves off Kerr black holes Journal Monthly Notices of the Royal Astronomical Society Date: Oct 01 2000 Abstract: Chandrasekhar separated the Dirac equation for spinning and massive particles in Kerr geometry into radial and angular parts. Here the authors solve the complete wave equation and find out how the Dirac wave scatters off Kerr black holes. The eigenfunctions, eigenvalues and reflection and transmission co-efficients are computed. The authors compare the solutions with several parameters to show how a spinning black hole recognizes the mass and energy of incoming waves. Journal volume: 317 Issue number: 4 Page number: 979-984 DOE sponsorship Publisher: Blackwell Science

Chamblin, A.; Hawking, S. W.; Reall, H. S. Brane-world black holes Journal Physical Review Date: Mar 15 2000 Abstract: Abstract available at publisher web site Journal volume: 61 Issue number: 6 Page number: 065007-065007-6 Publisher: American Physical Society

Chamblin, Andrew Capture of bulk geodesics by brane-world black holes Journal Classical and Quantum Gravity Date: Feb 07 2001 Abstract: P In this letter the authors study bulk time-like geodesics in the presence of a brane-world black hole, where the brane-world is a two-brane moving in a (3+1)-dimensional asymptotically adS 4 spacetime. The authors show that for a certain range of the parameters measuring the black hole mass and bulk cosmological constant, there exist stable time-like geodesics which orbit the black hole and remain bound close to the two-brane. Journal volume: 18 Issue number: 3 Page number: L17-L22 Publisher: Institute of Physics

Chamblin, Andrew; Csaki, Csaba; Erlich, Joshua; Hollowood, Timothy J. Black diamonds at brane junctions Journal

Physical Review Date: Aug 15 2000 Abstract: Abstract available at publisher web site Journal volume: 62 Issue number: 4 Page number: 044012-044012-7 Publisher: American Physical Society

Chamblin, Andrew; Csaki, Csaba; Erlich, Joshua; Hollowood, Timothy J. [Theory Division T-8, Los Alamos National Laboratory, Los Alamos, New Mexico 87545 (United States); Department of Physics, University of Wales Swansea, Swansea, SA2 8PP, (United Kingdom)] Black diamonds at brane junctions Journal Physical Review. Particles Fields Date: Aug 15 2000 Abstract: The authors discuss the properties of black holes in brane-world scenarios where our Universe is viewed as a four-dimensional sub-manifold of some higher-dimensional spacetime. The authors consider in detail such a model where four-dimensional spacetime lies at the junction of several domain walls in a higher dimensional anti-de Sitter spacetime. In this model there may be any number p of infinitely large extra dimensions transverse to the brane-world.

Chamblin, Andrew; Emparan, Roberto; Johnson, Clifford V.; Myers, Robert C. Holography, thermodynamics, and fluctuations of charged AdS black holes Journal Physical Review Date: Nov 15 1999 Abstract: Abstract available at publisher web site Journal volume: 60 Issue number: 10 Page number: 104026-104026-20 Publisher: American Physical Society

Chamblin, Andrew; Reall, Harvey S.; Shinkai, Hisa-aki; Shiromizu, Tetsuya Charged brane-world black holes Journal Physical Review Date: Mar 15 2001 Abstract: The authors study charged brane-world black holes in the model of Randall and Sundrum in which our universe is viewed as a domain wall in asymptotically anti--de Sitter space. Such black holes can carry two types of "charge," one arising from the bulk Weyl tensor and one from a gauge field trapped on the wall. The authors use a combination of analytical and numerical techniques to study how these black holes behave in the bulk.

Chamblin, Andrew; Reall, Harvey S.; Shinkai, Hisa-aki; Shiromizu, Tetsuya Charged brane-world black holes Journal Physical Review Date: Mar 15 2001 Abstract: Abstract available at publisher web site Journal volume: 63 Issue number: 6 Page number: 064015-064015-11 Publisher: American Physical Society

Chamseddine, A.; Ferrara, S.; Gibbons, G.W.; Kallosh, R. Enhancement of supersymmetry near a 5D black hole horizon Journal Physical Review, D (Particles Fields) Date: Mar 31 1997 Abstract: Geometric Killing spinors which exist on $AdS_{p+2} \times S^d - p - 2$ sometimes may be identified with supersymmetric Killing spinors. This explains the enhancement of unbroken supersymmetry near the p-brane horizon in d dimensions. The corresponding p-brane interpolates between two maximally supersymmetric vacua, at infinity and at the horizon. Journal volume: 55 Issue number: 6 Page number: 3647-3653 Subjects: Black Holes/supersymmetry; Supersymmetry; Spinors; Vacuum States; Geometry; Entropy; Many-Dimensional Calculations

Chattopadhyay, Indranil; Chakrabarti, Sandip K.; Fang, L. Z. Investigation of Radiative Outflows around Compact Objects Journal International Journal of Modern Physics Date: Feb 01 2000 Abstract: Winds and outflows form in active galaxies and in binary systems which are known to harbour compact objects such as black holes. Matter starting subsonically from a disc must be accelerated very close to the black hole in order to reach a velocity comparable to the velocity of light, which is actually observed. In the absence of magnetic fields, winds forming in inner regions of accretion discs could primarily be accelerated by radiations emitted from this region where centrifugal force is important. The authors study critical point behaviour of outflows in presence of this radiative acceleration.

Cherepashchuk, A.M. X-ray Nova Binary Systems Journal Space Science Reviews Date: Aug 01 2000 Abstract: The physical properties of X-ray novae as close binary systems are analysed. Observational data in X-ray, UV, optical, IR and radio ranges are summarized. Modern theoretical considerations of the problem of X-ray novae, taking into account some new ideas and results, are described. Properties of optical stars in X-ray novae are analysed. Data about the masses of black holes in X-ray binary systems are summarized. Possible evolutionary links between WR stars in close binary systems and X-ray novae are analysed. Journal volume: 93 Issue number: 3/4 Page number: 473-580 Rights: Kluwer Academic Publishers Publisher: Kluwer Academic Publishers

Chimento, Luis P.; Jakubi, Alejandro.S.; Pavón, Diego Stability of inflationary solutions driven by a changing dissipative fluid Journal Classical and Quantum Gravity Date: May 01 1999 Abstract: In this paper the second Lyapunov method is used to study the stability of the de Sitter phase of cosmic expansion when the source of the gravitational field is a viscous fluid. Different inflationary scenarios related to reheating and decay of mini-black holes into radiation are investigated using an effective fluid described by time-varying thermodynamical quantities. Journal volume: 16 Issue number: 5 Page number: 1625-1635 Publisher: Institute of Physics

Choptuik, Matthew W.; Hirschmann, Eric W.; Marsa, Robert L. New critical behavior in Einstein-Yang-Mills collapse Journal Physical Review Date: Dec 15 1999 Abstract: Abstract available at publisher web site Journal volume: 60 Issue number: 12 Page number: 124011-124011-9 Publisher: American Physical Society

Chou, W; Tajima, T. Dynamics of Plasma Close to the Horizon of a Schwarzschild Black Hole Journal Astrophysical Journal Date: Mar 31 1999 Abstract: General relativistic plasma dynamics relevant to the condition very close to a black hole event horizon is developed. The plasma is studied using the 3+1 paradigm of general relativistic magnetohydrodynamics. The equilibrium and dynamical solution of such a plasma in Rindler's coordinates are presented. Journal volume: 513 Issue number: 1 Page number: 401-408 Subjects: Schwarzschild Metric; Plasma Instability; Relativistic Plasma; Magnetic Fields; Hydrodynamics; Magnetohydrodynamics

Christodoulakis, T.; Diamandis, G. A.; Georgalas, B. C.; Vagenas, E. C. Casimir effect in 2D stringy black hole backgrounds Journal Physical Review Date: Dec 15 2001 Abstract: Abstract available at publisher web site Journal volume: 64 Issue number: 12 Page number: 124022-124022-12 Publisher: American Physical Society

Chrusciel, Piotr T. The classification of static vacuum spacetimes containing an asymptotically flat spacelike hypersurface with compact interior Journal Classical and Quantum Gravity Date: Mar 01 1999 Abstract: The authors prove non-existence of static, vacuum, appropriately regular, asymptotically flat black hole spacetimes with degenerate (not necessarily connected) components of the event horizon. This finishes the classification of static, vacuum, asymptotically flat domains of outer communication in an appropriate class of spacetimes, showing that the domains of outer communication of the Schwarzschild black holes exhaust the space of appropriately regular black hole exteriors. Journal volume: 16 Issue number: 3 Page number: 661-687 Publisher: Institute of Physics

Churazov, E.; Gilfanov, M.; Revnivtsev, M. Soft state of Cygnus X-1: stable disc and unstable corona Journal Monthly Notices of the Royal Astronomical Society Date: Mar 01 2001 Abstract: Two-component X-ray spectra (soft multicolour black-body and harder power law) are frequently observed from accreting black holes. These components are presumably associated with the different parts of the accretion flow (optically thick and optically thin respectively) in the vicinity of the compact source. Most of the aperiodic variability of the X-ray flux on the short time-scales is associated with the harder component. The authors suggest that drastically different amplitudes of variability of these two components are simply related to the very different viscous time-scales in the geometrically thin and geometrically thick parts of the accretion flow.

Clément, G.; Fabbri, A. The gravitating model in 2 + 1 dimensions: black hole solutions Journal Classical and Quantum Gravity Date: Feb 01 1999 Abstract: The authors derive and discuss black-hole solutions to the gravitating model in (2 + 1) dimensions. Three different kinds of static black holes are found. One of these resembles the static BTZ black hole, another is completely free of singularities, and the last type has the same Penrose diagram as the (3 + 1)-dimensional Schwarzschild black hole. The authors also construct static and dynamical multi-black-hole systems. Journal volume: 16 Issue number: 2 Page number: 323-341 Publisher: Institute of Physics

Clément, Gérard; Gal'tsov, Dmitri Bertotti-Robinson-type solutions to dilaton-axion gravity Journal Physical Review Date: Jun 15 2001 Abstract: Abstract available at publisher web site Journal volume: 63 Issue number: 12 Page number: 124011-124011-12 Publisher: American Physical Society

Clément, Gérard; Gal'tsov, Dmitri Solitons and black holes in Einstein-Born-Infeld-dilaton theory Journal Physical Review Date: Dec 15 2000 Abstract: Abstract available at publisher web site Journal volume: 62 Issue number: 12 Page number: 124013-124013-10 Publisher: American Physical Society

Clement, Gerard; Gal'tsov, Dmitri Bertotti-Robinson-type solutions to dilaton-axion gravity Journal Physical Review Date: Jun 15 2001 Abstract: The authors present a new solution to dilaton-axion gravity which looks like a rotating Bertotti-Robinson (BR) universe. It is supported by an homogeneous Maxwell field and a linear axion and can be obtained as a near-horizon limit of extremal rotating dilaton-axion black holes.

Cohen, A.G.; Kaplan, D.B.; Nelson, A.E. Effective Field Theory, Black Holes, and the Cosmological Constant Journal Physical Review Letters Date: Jun 01 1999 Abstract: Bekenstein has proposed the bound $S\{le\}\{pi\}M\{sup\ 2\}\{sub\ P\}L\{sup\ 2\}$ on the total entropy S in a volume $L\{sup\ 3\}$. This nonextensive scaling suggests that quantum field theory breaks down in large volume. To reconcile this breakdown with the success of local quantum field theory in describing observed particle phenomenology, the authors propose a relationship between UV and IR cutoffs such that an effective field theory should be a good description of nature.

Cohen, Andrew G.; Kaplan, David B.; Nelson, Ann E. Effective Field Theory, Black Holes, and the Cosmological Constant Journal Physical Review Letters Date: Jun 21 1999 Abstract: Abstract available at publisher web site Journal volume: 82 Issue number: 25 Page number: 4971-4974 Publisher: American Physical Society

Coleman Miller, M.; Hamilton, Douglas P. Production of intermediate-mass black holes in globular clusters Journal Monthly Notices of the Royal Astronomical Society Date: Jan 31 2002 Abstract: Abstract available at Blackwell Science. Journal volume: 330 Issue number: 1 Page number: 232-240 DOE sponsorship Publisher: Blackwell Science, Ltd

Collins, G.P. Quantum black holes are tied to D-branes and strings Journal Physics Today Date: Mar 31 1997 Abstract: Theorists have used string theory and nonperturbative objects known as D-branes to count the quantum microstates of black holes. This major step forward in our understanding of quantum gravity may resolve the longstanding "paradox" of information loss in black hole evaporation. Journal volume: 50 Issue number: 3 Page number: 19-22 Subjects: Black Holes/string models; General Relativity Theory; Quantum Mechanics; Entropy; Ground States; Supersymmetry; Quantum Gravity; Gravitational Radiation

Collins, G.P. Quantum black holes are tied to D-branes and strings Journal Physics Today Date: Mar 31 1997 Abstract: Theorists have used string theory and nonperturbative objects known as D-branes to count the quantum microstates of black holes. This major step forward in our understanding of quantum gravity may resolve the longstanding "paradox" of information loss in black hole evaporation. Journal volume: 50 Issue number: 3 Page number: 19-22 Subjects: Black Holes/string models; General Relativity Theory; Quantum Mechanics; Entropy; Ground States; Supersymmetry; Quantum Gravity; Gravitational Radiation

Collins, G.P. Quantum black holes are tied to D-branes and strings Journal Physics Today Date: Mar 31 1997 Abstract: Theorists have used string theory and nonperturbative objects known as D-branes to count the quantum microstates of black holes. This major step forward in our understanding of quantum gravity may resolve the longstanding "paradox" of information loss in black hole evaporation. Journal volume: 50 Page number: 19-22 SubjectsEntropy; General Relativity Theory; Gravitational Radiation; Ground States; Quantum Gravity; Quantum Mechanics; Supersymmetry; Energy Levels; Extended Particle Model; Field Theories; Mathematical Models; Quantum Field Theory; Quark Model; Radiations

Constable, Neil R. Entropy of 4D black holes and the enhançon Journal Physical Review Date: Nov 15 2001 Abstract: Abstract available at publisher web site Journal volume: 64 Issue number: 10 Page number: 104004-104004-13 Publisher: American Physical Society

Corbin, Michael R. On the Role of Minor Galaxy Mergers in the Formation of Active Galactic Nuclei Journal The Astrophysical Journal Date: Jun 01 2000 Abstract: The large-scale (~100 kpc) environments of Seyfert galaxies are not significantly different from those of non-Seyfert galaxies. In the context of the interaction model of the formation of active galactic nuclei (AGNs), it has thus been proposed that AGNs form via "minor mergers" of large disk galaxies with smaller companions. The authors test this hypothesis by comparing the nuclear spectra of 105 bright nearby galaxies with measurements of their R- or r-band morphological asymmetries at three successive radii.

Corbin, Michael R. On the Role of Minor Galaxy Mergers in the Formation of Active Galactic Nuclei Journal The Astrophysical Journal Date: Jun 01 2000 Abstract: The large-scale (~100 kpc) environments of Seyfert galaxies are not significantly different from those of non-Seyfert galaxies. In the context of the interaction model of the formation of active galactic nuclei (AGNs), it has thus been proposed that AGNs form via "minor mergers" of large disk galaxies with smaller companions. The authors test this hypothesis by comparing the nuclear spectra of 105 bright nearby galaxies with measurements of their R- or r-band morphological asymmetries at three successive radii. The authors find no significant differences between these asymmetries among the 13 Seyfert galaxies in the sample and galaxies having other nuclear spectral types (absorption, H ii region-like, LINER), nor is there strong qualitative evidence that such mergers have occurred among any of the Seyfert galaxies or liners.

Corichi, Alejandro; Nucamendi, Ulises; Sudarsky, Daniel Einstein-Yang-Mills isolated horizons: Phase space, mechanics, hair, and conjectures Journal Physical Review Date: Aug 15 2000 Abstract: Abstract available at publisher web site Journal volume: 62 Issue number: 4 Page number: 044046-044046-19 Publisher: American Physical Society

Corichi, Alejandro; Nucamendi, Ulises; Sudarsky, Daniel Mass formula for Einstein-Yang-Mills solitons Journal Physical Review Date: Nov 15 2001 Abstract: Abstract available at publisher web site Journal volume: 64 Issue number: 10 Page number: 107501-107501-4 Publisher: American Physical Society

Corichi, Alejandro; Sudarsky, Daniel Mass of colored black holes Journal Physical Review Date: May 15 2000 Abstract: Abstract available at publisher web site Journal volume: 61 Issue number: 10 Page number: 101501-101501-4 Publisher: American Physical Society

Costa, M.S.; Cvetic, M. Nonthreshold D-brane bound states and black holes with nonzero entropy Journal Physical Review, D (Particles Fields) Date: Oct 31 1997 Abstract: The authors start with Bogomol'nyi-Prasad-Sommerfield- (BPS) saturated configurations of two (orthogonally) intersecting M-branes and use the electromagnetic duality or dimensional reduction along a boost, in order to obtain new p-brane bound states. In the first case the resulting configurations are interpreted as BPS-saturated nonthreshold bound states of intersecting p-branes, and in the second case as p-branes intersecting at angles and their duals. As a by-product the authors deduce the enhancement of supersymmetry as the angle approaches zero. Journal volume: 56 Issue number: 8 Page number: 4834-4843 Subjects: Black Holes/entropy; Solitons/bound state; Entropy; Solitons; Duality; Supersymmetry; Unified Gauge Models

Coule, D.H. Does brane cosmology have realistic principles? Journal Classical and Quantum Gravity Date: Oct 21 2001 Abstract: P The maximal symmetry, or perfect cosmological principle (PCP), that prevents AdS-type spaces from degenerating into anti-inflationary collapse is argued to be unphysical. For example, the simple requirement that brane-bulk models should be the result of having evolved from even more energetic string phenomena picks out a preferred time direction. The authors question whether quantum cosmological reasoning can be applied in any meaningful way to obtain, what are essentially, classical constructs.

Crisóstomo, Juan; Troncoso, Ricardo; Zanelli, Jorge Black hole scan Journal Physical Review Date: Oct 15 2000 Abstract: Abstract available at publisher web site Journal volume: 62 Issue number: 8 Page number: 084013-084013-14 Publisher: American Physical Society

Crispino, Luis C. B.; Higuchi, Atsushi; Matsas, George E. A. Quantization of the electromagnetic field outside static black holes and its application to low-energy phenomena Journal Physical Review Date: Jun 15 2001 Abstract: The authors discuss the Gupta-Bleuler quantization of the free electromagnetic field outside static black holes in the Boulware vacuum. The authors use a gauge which reduces to the Feynman gauge in Minkowski spacetime. The authors also discuss its relation with gauges used previously.

Croce, R. P.; Demma, Th.; Pierro, V.; Pinto, I. M.; Churches, D.; Sathyaprakash, B. S. Gravitational wave chirp search: Economization of post-Newtonian matched filter bank via cardinal interpolation Journal Physical Review Date: Dec 15 2000 Abstract: Abstract available at publisher web site Journal volume: 62 Issue number: 12 Page number: 121101-121101-5 Publisher: American Physical Society

Cruz, J.; Fabbri, A.; Navarro, D. J.; Navarro-Salas, J. Integrable models and degenerate horizons in two-dimensional gravity Journal Physical Review Date: Jan 15 2000 Abstract: Abstract available at publisher web site Journal volume: 61 Issue number: 2 Page number: 024011-024011-9 Publisher: American Physical Society

Cui, Wei; Zhang, Shuang Nan; Chen, Wan Phase Lag and Coherence Function of X-Ray Emission from Black Hole Candidate XTE J1550-564 Journal The Astrophysical

Journal Date: Mar 01 2000 Abstract: The authors report the results from measuring the phase lag and coherence function of X-ray emission from black hole candidate XTE J1550-564. These temporal X-ray properties have been recognized to be increasingly important in providing important diagnostics of the dynamics of accretion flows around black holes. For XTE J1550-564, the authors found significant hard lag—the X-ray variability in high-energy bands lags behind that in low-energy bands—associated both with broadband variability and quasi-periodic oscillation (QPO). However, the situation is more complicated for the QPO: while hard lag was measured for the first harmonic of the signal, the fundamental component showed significant soft lag.

Custódio, P. S.; Horvath, J. E. Bounds on the cosmological abundance of primordial black holes from diffuse sky brightness: Single mass spectra Journal Physical Review Date: Jan 15 2002 Abstract: Abstract available at publisher web site Journal volume: 65 Issue number: 2 Page number: 024023-024023-8 Publisher: American Physical Society

Custódio, P. S.; Horvath, J. E. Dynamics of black hole motion Journal Physical Review Date: Oct 15 1999 Abstract: Abstract available at publisher web site Journal volume: 60 Issue number: 8 Page number: 083002-083002-9 Publisher: American Physical Society

Cveti, Mirjam; Larsen, Finn Statistical Entropy of Four-Dimensional Rotating Black Holes from Near-Horizon Geometry Journal Physical Review Letters Date: Jan 18 1999 Abstract: Abstract available at publisher web site Journal volume: 82 Issue number: 3 Page number: 484-487 Publisher: American Physical Society

Cvetic, M; Larsen, F. General rotating black holes in string theory: Greybody factors and event horizons Journal Physical Review, D (Particles Fields) Date: Oct 31 1997 Abstract: The authors derive the wave equation for a minimally coupled scalar field in the background of a general rotating five-dimensional black hole. It is written in a form that involves two types of thermodynamic variables, defined at the inner and outer event horizon, respectively. The authors model the microscopic structure as an effective string theory, with the thermodynamic properties of the left- and right-moving excitations related to those of the horizons. Journal volume: 56 Issue number: 8 Page number: 4994-5007 Subjects: Rotation; Wave Equations; Scalar Fields; Thermodynamic Properties; Emission Spectra; Cross Sections

Cvetic, M; Larsen, F. Greybody factors for black holes in four dimensions: Particles with spin Journal Physical Review, D (Particles Fields) Date: May 31 1998 Abstract: The authors compute the emission spectrum of minimally coupled particles with spin that are Hawking radiated from four-dimensional black holes in string theory. For a range of the black hole parameters the result has a product structure that may be interpreted in terms of the respective right- and left-moving thermal correlation functions of an effective string model. Journal volume: 57 Issue number: 10 Page number: 6297-6310 Subjects: Stellar Radiation; Four-Dimensional Calculations; Correlation Functions; Wave Functions

Cvetic, M; Larsen, F.; Cvetic, M. Statistical Entropy of Four-Dimensional Rotating Black Holes from Near-Horizon Geometry Journal Physical Review Letters Date: Jan 31 1999 Abstract: The authors show that a class of four-dimensional rotating black holes allow five-dimensional embeddings as black rotating strings. Their near-horizon geometry factorizes locally as a product of the three-dimensional anti–de Sitter space-time and a two-dimensional sphere (AdS{sub 3}{times}S{sup 2}), with angular momentum encoded in the global space-time structure. Journal volume: 82 Issue number: 3 Page number: 484-487 Subjects: Rotation; Space-Time; String Models; Angular Momentum

Czerny, B.; Nikolstrok;ajuk, M.; Piasecki, M.; Kuraszkiewicz, J. Black hole masses from power density spectra: determinations and consequences Journal Monthly Notices of the Royal Astronomical Society Date: Aug 01 2001 Abstract: The authors analyse the scaling of the X-ray power density spectra with the mass of the black hole in the examples of Cyg X-1 and the Seyfert 1 galaxy NGC 5548. The authors show that the high-frequency tail of the power density spectrum can be successfully used for the determination of the black hole mass. The authors determine the masses of the black holes in six broad-line Seyfert 1 galaxies, five narrow-line Seyfert 1 galaxies and two quasi-stellar objects (QSOs) using the available power density spectra.

Dain, Sergio Initial data for a head-on collision of two Kerr-like black holes with close limit Journal Physical Review Date: Dec 15 2001 Abstract: Abstract available at publisher web site Journal volume: 64 Issue number: 12 Page number: 124002-124002-7 Publisher: American Physical Society

Dain, Sergio Initial Data for Two Kerr-like Black Holes Journal Physical Review Letters Date: Sep 17 2001 Abstract: Abstract available at publisher web site Journal volume: 87 Issue number: 12 Page number: 121102-121102-4 Publisher: American Physical Society

Damour, Thibault Coalescence of two spinning black holes: An effective one-body approach Journal Physical Review Date: Dec 15 2001 Abstract: Abstract available at publisher web site Journal volume: 64 Issue number: 12 Page number: 124013-124013-22 Publisher: American Physical Society

Daniel, J.; Tajima, T. Electromagnetic waves in a strong Schwarzschild plasma Journal Physical Review, Date: Apr 30 1997 Abstract: The physics of high-frequency electromagnetic waves in a general relativistic plasma with the Schwarzschild metric is studied. Based on the 3+1 formalism, the authors conformalize Maxwell's equations. The derived dispersion relations for waves in the plasma contain the lapse function in the plasma parameters such as in the plasma frequency and cyclotron frequency, but otherwise look "flat." Journal volume: 558 Page number: 5193-5204 Subject keyword: Black Holes; Dispersion Relations; Electromagnetic Fields; Electron-Positron Interactions; General Relativity Theory; Maxwell Equations; Plasma Simulation; Plasma Waves; Relativistic Plasma; Schwarzschild Metric; Spectroscopy; Unified-Field Theories; Wave Equations; Differential Equations; Equations; Field Theories; Interactions; Lepton-Lepton Interactions; Metrics; Partial Differential Equations; Particle Interactions; Plasma

Daniel, J; Tajima, T. Outbursts from a Black Hole via Alfv{acute e}n Wave to Electromagnetic Wave Mode Conversion Journal Astrophysical Journal Date: May 31 1998 Abstract: A new mechanism for outbursts from a black hole is proposed. A recent work on general relativistic plasma equilibria around a black hole has shown the possibility of equilibrium presence of matter and magnetic fields in the neighborhood of the event horizon, even where the corpuscular equilibrium is not allowed ($R \{lt\} 3R\{sub\ s\}$, where $R\{sub\ s\}$ is the Schwarzschild radius). A large-amplitude Alfv{acute e}n pulse in the black hole electron-positron atmosphere that propagates away from the hole into lower magnetic field regions can experience resonance and mode-convert itself into a large-amplitude electromagnetic (EM) pulse. Journal volume: 498 Issue number: 1 Page number: 296-306 Subjects: Photoemission; Energy Transfer; Relativistic Plasma; Plasma Waves; Mode Conversion; Energy Spectra

Das, S.R.; Mathur, S.D.; Ramadevi, P. Hawking radiation from four-dimensional Schwarzschild black holes in M theory Journal Physical Review, D (Particles Fields) Date: Apr 30 1999 Abstract: Recently a method has been developed for relating four dimensional Schwarzschild black holes in M theory to near-extremal black holes in string theory with four charges, using suitably defined "boosts" and T dualities. The authors show that this method can be extended to obtain the emission rate of low energy massless scalars for the four dimensional Schwarzschild hole from the microscopic picture of radiation from the near extremal hole. Journal volume: 59 Issue number: 8 Page number: 084001 Subjects: Schwarzschild Metric; String Models;

Das, S; Dasgupta, A; Sarkar, T. High energy effects on D-brane and black hole emission rates Journal Physical Review, D (Particles Fields) Date: Jun 30 1997 Abstract: The authors study the emission of scalar particles from a class of near-extremal five-dimensional black holes and the corresponding D-brane configuration at high energies. The authors show that the distribution functions and the black hole greybody factors are modified in the high energy tail of the Hawking spectrum in such a way that the emission rates exactly match. Journal volume: 55 Issue number: 12 Page number: 7693-7700 Subjects: Emission; Solitons; Many-Dimensional Calculations; Spectra; Scalar Fields

Das, S; Majumdar, P. Eikonal particle scattering and dilaton gravity Journal Physical Review, D (Particles Fields) Date: Feb 28 1997 Abstract: Approximating light charged pointlike particles in terms of (nonextremal) dilatonic black holes is shown to lead to certain pathologies in Planckian scattering in the eikonal approximation, which are traced to the presence of a (naked) curvature singularity in the metric of these black holes. The existence of such pathologies is confirmed by analyzing the problem in an "external metric" formulation where an ultrarelativistic point particle scatters off a dilatonic black hole geometry at large impact parameters. Journal volume: 55 Issue number: 4 Page number: 2090-2098 Subject Keyword: Quantum Gravity; Scattering; Singularity; Geometry; Impact Parameter; Light Cone

Das, Saurya; Gegenberg, J.; Husain, V.; Ahluwalia, D. V. Can Black Holes Decay to Naked Singularities & quest; Journal International Journal of Modern Physics A Date: Dec 01 2001 Abstract: The authors investigate thermodynamic properties of two types of asymptotically anti-de Sitter spacetimes & colon; black holes and singular scalar field spacetimes. The authors describe the possibility that thermodynamic phase transitions can transform one spacetime into another, suggesting that black holes can radiate to naked singularities. Journal volume: 10 Issue number: 06 Rights: World Scientific Publishing Company Publisher: World Scientific Publishing Company

Das, Saurya; Ghosh, Amit; Mitra, P. Statistical entropy of Schwarzschild black strings and black holes Journal Physical Review Date: Jan 15 2001 Abstract: Abstract available at publisher web site Journal volume: 63 Issue number: 2 Page number: 024023-024023-4 Publisher: American Physical Society

Das, Tapas K. On some transonic aspects of general relativistic spherical accretion on to Schwarzschild black holes Journal Monthly Notices of the Royal Astronomical Society Date: Jan 31 2002 Abstract: Abstract available at Blackwell Science. Journal volume: 330 Issue number: 3 Page number: 563-566 DOE sponsorship. Blackwell Science, Ltd

Das, Tapas K.; Chakrabarti, Sandip.K. Mass outflow rate from accretion discs around compact objects Journal Classical and Quantum Gravity Date: Dec 01 1999 Abstract: The authors compute mass outflow rates from accretion discs around compact objects, such as neutron stars and black holes. These computations are done using combinations of exact transonic inflow and outflow solutions which may or may not form standing shock waves. Assuming that the bulk of the outflow is from the effective boundary layers of these objects, the authors find that the ratio of the outflow and inflow rates varies anywhere from a few per cent to even close to a 100% (i.e. close to the disc evacuation case) depending on the initial parameters of the disc, the degree of compression of matter near the centrifugal barrier, and the polytropic index of the flow.

de Alwis, S.P; Sato, K. Radiation from a class of string theoretic black holes Journal Physical Review, D (Particles Fields) Date: May 31 1997 Abstract: The emission of a scalar with low energy {omega}, from a D-(4{le}D{le}8) dimensional black hole with n charges, is studied in both string and semiclassical calculations. In the lowest order in {omega}, the weak coupling string and semiclassical calculations agree provided that the Bekenstein-Hawking formula is valid and the effective central charge c{sub eff}=6 for any D. Journal volume: 55 Issue number: 10 Page number: 6181-6188 Subjects: Radiations; Semiclassical Approximation; Coupling

de Moura, Alessandro P. S.; Letelier, Patricio S. Scattering map for two black holes Journal Physical Review E Date: Oct 31 2000 Abstract: Abstract available at publisher web site Journal volume: 62 Issue number: 4 Page number: 4784-4791 Publisher: American Physical Society

Deane, J.R.; Trentham, Neil On the lack of cold dust in IRAS P09104+4109 and IRAS F15307+3252: their spectral energy distributions and implications for finding dusty AGNs at high redshift Journal Monthly Notices of the Royal Astronomical Society Date: Oct 01 2001 Abstract: The authors present upper limits on the 850- m and 450- m fluxes of the warm hyperluminous (bolometric luminosity &formmu1; galaxies IRAS P09104+4109 &formmu2; and IRAS F15307+3252 &formmu3;, derived from measurements using the SCUBA bolometer array on the James Clerk Maxwell Telescope. Hot luminous infrared sources like these are thought to differ from more normal cold ultraluminous infrared &formmu4; galaxies in that they derive most of their bolometric luminosities from dusty active galactic nuclei (AGNs) as opposed to starbursts. Such hot, dusty AGNs at high redshift are thought to be responsible for much of the mass accretion of the Universe that is in turn responsible for the formation of the supermassive black holes seen in the centres of local galaxies.

Degura, Yoshitaka; Shiraishi, Kiyoshi Effective field theory of slowly moving 'extreme black holes' Journal Classical and Quantum Gravity Date: Oct 07 2000 Abstract: P The authors consider the non-relativistic effective field theory of 'extreme black holes' in the Einstein-Maxwell-dilaton theory with an arbitrary dilaton coupling. The authors investigate the finite-temperature behaviour of a gas of 'extreme black holes' using the effective theory. The total energy of the classical many-body system is also derived. Journal volume: 17 Issue number: 19 Page number: 4031-4050 Publisher: Institute of Physics

Dehghani, M. H.; Mann, R. B. Quasilocal thermodynamics of Kerr and Kerr--anti-de Sitter spacetimes and the AdS/CFT correspondence Journal Physical Review Date: Aug 15 2001 Abstract: The authors consider the quasilocal thermodynamics of rotating black holes in asymptotically flat and asymptotically anti--de Sitter (AdS) spacetimes. Using the minimal number of intrinsic boundary counterterms inspired by the AdS/conformal field theory correspondence, the authors find that the authors are able to carry out an analysis of the thermodynamics of these black holes for virtually all possible values of the rotation parameter and cosmological constant that leave the quasilocal boundary well defined, going well beyond what is possible with background subtraction methods.

Dereli, T.; Sariog˘lu, O". Supersymmetric solutions to topologically massive gravity and black holes in three dimensions Journal Physical Review Date: Jul 15 2001 Abstract: Abstract available at publisher web site Journal volume: 64 Issue number: 2 Page number: 027501-027501-4 American Physical Society

Dereli, Tekin; Obukhov, Yuri N. Massless scalar fields and topological black holes Journal Physical Review Date: Apr 15 2000 Abstract: Abstract available at publisher web site Journal volume: 61 Issue number: 8 Page number: 084015-084015-7 Publisher: American Physical Society

Dias, Oscar J. C.; Lemos, José P. S. Static and rotating electrically charged black holes in three-dimensional Brans-Dicke gravity theories Journal Physical Review Date: Sep 15 2001 Abstract: Abstract available at publisher web site Journal volume: 64 Issue number: 6 Page number: 064001-064001-14 Publisher: American Physical Society

Dimopoulos, Savas; Landsberg, Greg Black Holes at the Large Hadron Collider Journal Physical Review Letters Date: Oct 15 2001 Abstract: If the scale of quantum gravity is near TeV, the CERN Large Hadron Collider will be producing one black hole (BH) about every second. The decays of the BHs into the final states with prompt, hard photons, electrons, or muons provide a clean signature with low background. The correlation between the BH mass and its temperature, deduced from the energy spectrum of the decay products, can test Hawking's evaporation law and determine the number of large new dimensions and the scale of quantum gravity. Journal volume: 87 Issue number: 16 Page number: 161602-161602-4 Publisher: The American Physical Society

Discovery of a massive equatorial torus in the η Carinae stellar system Journal Nature Date: Dec 02 1999 Abstract: The enigmatic object η Carinae is believed to represent an important, but short-lived, unstable phase in the life of the most massive stars, occurring shortly before they explode as supernovae or collapse directly to black holes. The putative binary. The physical mechanisms responsible for the outburst and the bipolar geometry are not understood. Here the authors report infrared observations (spectroscopy and imaging) that reveal the presence of about 15 solar masses of material, located in an equatorial torus. The massive torus may have been created through highly non-conservative mass transfer, which removed the entire envelope of one of the stars, leaving an unstable core that erupted in the nineteenth century. The collision of the erupted material with the pre-existing torus provides a natural explanation for the bipolar shape of the nebula. Journal volume: 402 Issue number: 6761 Page number: 502-504 Rights: 1999 Macmillan Publishers Ltd. Publisher: Nature Publishing Group - Nature

Dowker, F.; Kastor, D; Traschen, J. U duality, D-branes, and black hole emission rates: Agreements and disagreements Journal Physical Review, D (Particles Fields) Date: Dec 31 1998 Abstract: An expression for the spacetime absorption coefficient of a scalar field in a five-dimensional, near-extremal black hole background is derived, which has the same form as that presented by Maldacena and Strominger, but is valid over a larger, U-duality invariant region of parameter space and in general disagrees with the corresponding D-brane result. The authors develop an argument, based on D-brane thermodynamics, which specifies the range of parameters over which agreement should be expected. Journal volume: 58 Issue number: 12 Page number: 124025 Subjects: Black Holes; String Models; Extended Particle Model; Thermodynamics; Space-Time

Dreyer, Olaf; Ghosh, Amit; Wisniewski, Jacek Black hole entropy calculations based on symmetries Journal Classical and Quantum Gravity Date: May 21 2001 Abstract: P Symmetry-based approaches to the black hole entropy problem have a number of attractive features; in particular, they are very general and do not depend on the details of the quantization method. However, the authors point out that, of the two available approaches, one faces conceptual problems (also emphasized by others), while the second contains certain technical flaws. The authors correct these errors and, within the new, improved scheme, calculate the entropy of three-dimensional black holes. The authors find that, while the new symmetry vector fields are well defined on the 'stretched horizon', and lead to well defined Hamiltonians satisfying the expected Lie algebra, they fail to admit a well defined limit to the horizon. This suggests that, although the formal calculation can be carried out at the classical level, its real, conceptual origin probably lies in the quantum theory. Journal volume: 18 Issue number: 10 Page number: 1929-1938 Publisher: Institute of Physics

Duff, M.J.; editor The World in Eleven Dimensions Journal Classical and Quantum Gravity Date: Mar 21 2000 Abstract: The first superstring revolution, in 1984, was responsible for eliminating 11-dimensional supergravity from the list of candidates for the ultimate theory of nature. The return of 11-dimensional supergravity, in 1995, in a new guise called M-theory in the second superstring revolution is more dramatic than its exit 11 years ago. This is because the duality relationship between all the known string theories is incomplete without M-theory. Though this new avatar of 11-dimensional supergravity entered as a missing piece in the jigsaw puzzle of string dualities, it quickly established its special status by demonstrating that most of the string theories including the chiral ones can be derived from it in some specific limit.

Dunning-Davies, J. Undergraduate thermodynamics and black holes Journal European Journal of Physics Date: Jul 01 1997 Abstract: It is noted that certain well known formulae of thermodynamics are no longer valid when systems whose entropies are not extensive (that is, not homogeneous of degree one) are under consideration. Attention is drawn to the fact that the accepted form for the entropy of a black hole falls into this category and so great care must be exercised when using black holes to generate added interest in undergraduate thermodynamics courses.

Dvali, Gia; Gabadadze, Gregory; Kolanovic´, Marko; Nitti, Francesco Scales of gravity Journal Physical Review Date: Jan 15 2002 Abstract: Abstract available at publisher web site Journal volume: 65 Issue number: 2 Page number: 024031-024031-27 Publisher: American Physical Society

Dzhunushaliev, V.; Ahluwalia, D. V. Matching Condition on the Event Horizon and Holography Principle Journal International Journal of Modern Physics Date: Oct 01 2000 Abstract: It is shown that the event horizon of 4D black hole or $ds^2=0$ surfaces of multidimensional wormhole-like solutions reduce the amount of information necessary for determining the whole spacetime and hence satisfy the Holography principle. This leads to the fact that by matching two metrics on a $ds^2=0$ surface (an event horizon for 4D black holes) the authors can match only the metric components but not their derivatives. For example, this allows us to obtain a composite wormhole inserting a 5D wormhole-like flux tube between two Reissner-Nordström black holes and matching them on the event horizon. Using the Holography principle, the entropy of a black hole from the algorithm theory is obtained. Journal volume: 09 Issue number: 05 Rights: World Scientific Publishing Company Publisher: World Scientific Publishing Company

Easther, Richard; Parry, Matthew Gravity, parametric resonance, and chaotic inflation Journal Physical Review Date: Nov 15 2000 Abstract: The authors investigate the possibility that nonlinear gravitational effects influence the preheating era after inflation. Our work is based on numerical solutions of the inhomogeneous Einstein field equations, and is free of perturbative approximations. The one restriction the authors impose is to limit the in homogeneity to a single spatial direction. The authors compare our results to perturbative calculations and to solutions of the nonlinear field equations in a rigid (unperturbed) spacetime, in order to isolate gravitational phenomena.

Easther, Richard; Parry, Matthew Gravity, parametric resonance, and chaotic inflation

Journal Physical Review Date: Nov 15 2000 Abstract: Abstract available at publisher web site Journal volume: 62 Issue number: 10 Page number: 103503-103503-15 Publisher: American Physical Society

Ellis, J.; Mavromatos, N.E. High-energy QCD as a topological field theory Journal The European Physical Journal C Date: Mar 01 1999 Abstract: The authors propose an identification of the conformal field theory underlying Lipatov's spin-chain model of high-energy scattering in perturbative QCD.Publisher: Springer-Verlag New York, Inc.

Emparan, Roberto Black diholes Journal Physical Review Date: May 15 2000 Abstract: Abstract available at publisher web site Journal volume: 61 Issue number: 10 Page number: 104009-104009-7 Publisher: American Physical Society

Emparan, Roberto; Gregory, Ruth; Santos, Caroline Black holes on thick branes Journal Physical Review Date: May 15 2001 Abstract: Abstract available at publisher web site Journal volume: 63 Issue number: 10 Page number: 104022-104022-12 Publisher: American Physical Society

Emparan, Roberto; Horowitz, Gary T.; Myers, Robert C. Black Holes Radiate Mainly on the Brane Journal Physical Review Letters Date: Jul 17 2000 Abstract: Abstract available at publisher web site Journal volume: 85 Issue number: 3 Page number: 499-502 Publisher: American Physical Society

Emparan, Roberto; Johnson, Clifford V.; Myers, Robert C. Surface terms as counterterms in the AdS-CFT correspondence Journal Physical Review Date: Nov 15 1999 Abstract: Abstract available at publisher web site Journal volume: 60 Issue number: 10 Page number: 104001-104001-14 Publisher: American Physical Society

Enderlein, J. A heuristic way of obtaining the Kerr metric Journal American Journal of Physics Date: Sep 30 1997 Abstract: An intuitive, straightforward way of finding the metric of a rotating black hole is presented, based on the algebra of differential forms. The representation obtained for the metric displays a simplicity which is not obvious in the usual Boyer–Lindquist coordinates. Journal volume: 65 Issue number: 9 Page number: 897-902 Subjects: Black Holes/kerr metric; Kerr Metric/algebra; Kerr Metric/analytical solution; Rotation; Algebra; General Relativity Theory

Ergma; J., M. What can the authors learn from the surface chemical composition of the optical companions of Soft X-ray transients? Journal Astronomy and Astrophysics Date: May 11 2001 Abstract: Several evolutionary sequences with low-mass secondaries and black hole accretors are calculated. The angular momentum losses due to magnetic braking and gravitational wave radiation are included. Using full nuclear networks (p-p and CNO cycles) the authors follow carefully the evolution of the surface composition of the secondary star. The authors find that the surface chemical composition of the secondary star may give additional information whichhelps to understand the formation of soft X-ray transients with black holes as accretors.

Etesi, Gá A Global Uniqueness Theorem for Stationary Black Holes Journal Communications in Mathematical Physics Date: Aug 01 1998 Abstract: Abstract: A global uniqueness theorem for stationary black holes is proved as a direct consequence of the Topological Censorship Theorem and the topological classification of compact, simply connected four-manifolds. Publisher: Springer-Verlag New York, Inc.

F., B.; Meyer-Hofmeister, E. Truncation of geometrically thin disks around massive blackholes in galactic nuclei Journal Astronomy and Astrophysics Date: Mar 16 2001 Abstract: The concept of an advection-dominated accretion flow (ADAF), with or without wind loss, was used to describe the spectra of the galactic center source Sgr$\rm{A}^$, low-luminosity AGN and nuclei of elliptical galaxies (including M 87). The spectral fits of various authors show that the transition from the geometrically thin disk to the hot flow occurs at quite different distances, apparently not uniquely related to the mass flow rate in the disk. The authors compare these results with the results of theoretical modeling where the authors determine the truncation of the thin outer disk from the efficiency of mass evaporation.

Fabbri, A. Cauchy horizon stability in two-dimensional 'accelerated' black holes Journal Classical and Quantum Gravity Date: Feb 01 1998 Abstract: In this paper the authors analyse the stability of black hole Cauchy horizons arising in a class of two-dimensional dilaton gravity theories that describes 'accelerated' black holes. It is shown that, due to the characteristic asymptotic Rindler form of the metric or to the presence of an acceleration horizon, time-dependent gravitational perturbations generated in the external region do not necessarily blow up when propagated along the Cauchy horizon. There exists, in fact, a region of non-zero measure in the space of the parameters characterizing the solutions such that mass inflation is avoided and the spacetime geometry remains regular on this surface. Despite this fact, however, quantum backreaction seems to produce a scalar curvature singularity there. Journal volume: 15 Issue number: 2 Page number: 373-387 Publisher: Institute of Physics

Fabbri, Alessandro; Navarro, Diego J.; Navarro-Salas, José Evaporation of Near-Extremal Reissner-Nordström Black Holes Journal Physical Review Letters Date: Sep 18 2000 Abstract: Abstract available at publisher web site Journal volume: 85 Issue number: 12 Page number: 2434-2437 Publisher: American Physical Society

Fabian, A. C. The obscured growth of massive black holes Journal Monthly Notices of the Royal Astronomical Society Date: Oct 01 1999 Abstract: The mass density of massive black holes observed locally is consistent with the hard X-ray background provided that most of the radiation produced during their growth was absorbed by surrounding gas. A simple model is proposed here for the formation of galaxy bulges and central black holes in which young spheroidal galaxies have a significant distributed component of cold dusty clouds, which accounts for the absorption. The central accreting black hole is assumed to emit both a quasar-like spectrum, which is absorbed by the surrounding gas, and a slow wind. The power in both is less than the Eddington limit for the black hole.

Fabian, A.C.; Wilman, R.J.; Crawford, C.S. On the detectability of distant Compton-thick obscured quasars Journal Monthly Notices of the Royal Astronomical Society Date: Jan 01 2002 Abstract: Chandra and XMM-Newton have resolved the &formmu1; X-ray background (XRB) into point sources. Many of the fainter sources are obscured active galactic nuclei (AGN) with column densities in the range of &formmu2; some of which have quasar-like luminosities. Here, the authors consider whether Compton-thick quasars are detectable by Chandra and XMM-Newton by their direct (i.e. not scattered) X-ray emission. Detectability is optimized if the objects individually have a high luminosity and high redshift, so that the direct emission has a significant flux in the observed band. Using a simple galaxy formation model incorporating accreting black holes, in which quasars build most of their mass in a Compton-thick manner before expelling the obscuring matter, the authors predict that moderately deep 100-ks Chandra and XMM-Newton exposures may contain a handful of detectable Compton-thick quasars.

Fender, R.P.; Kuulkers, E. On the peak radio and X-ray emission from neutron star and black hole candidate X-ray transients

Journal Monthly Notices of the Royal Astronomical Society Date: Jul 01 2001 Abstract: The authors have compiled and analysed reports from the literature of (quasi-)simultaneous observations of X-ray transients at radio and X-ray wavelengths and compared them with each other and with more unusual radio-bright sources such as Cygnus X-3, GRS 1915+105 and Circinus X-1.

Feng, Jonathan L.; Shapere, Alfred D. Black Hole Production by Cosmic Rays Journal Physical Review Letters Date: Jan 14 2002 Abstract: Abstract available at publisher web site Journal volume: 88 Issue number: 2 Page number: 021303-021303-4 Publisher: American Physical Society

Ferrara, Sergio; Sokatchev, Emery Conformal Superfields and BPS States in AdS4/7 Geometries Journal International Journal of Modern Physics B Date: Sep 20 2000 Abstract: The authors carry out a general analysis of the representations of thesuperconformal algebras $OSp(8/4, R)$ andOSp(8/2N) in terms of harmonic superspace. The authors present a construction of their highest-weight UIR's by multiplication of thedifferent types of massless conformal superfields("supersingletons").Particular attention is paid to the so-called "short multiplets".Representations undergoing shortening have "protected dimension"and may correspond to BPS states in the dual supergravity theoryin anti-de Sitter space.These results are relevant for the classification of multitraceoperators in boundary conformally invariant theories as well as forthe classification of AdS black holes preserving different fractions of supersymmetry. Journal volume: 14 Issue number: 22_23 Rights: World Scientific Publishing Company Publisher: World Scientific Publishing Company

Ferrara, Sergio; Sokatchev, Emery Representations of Superconformal Algebras in the AdS 7/4/CFT 6/3 Correspondence Journal International Journal of Modern Physics A Date: Feb 01 2001 Abstract: The authors perform a general analysis of representations of the superconformalalgebras $OSp(8/4, R)$ and $OSp(8/2N)$ in harmonicsuperspace. The authors present a construction of their highest-weight UIR's bymultiplication of the different types of massless conformalsuperfields ("supersingletons"). In particular, all "shortmultiplets" are classified. Representations undergoing shorteninghave "protected dimension" and may correspond to BPS states in thedual supergravity theory in anti-de Sitter space. These results arerelevant for the classification of multitrace operators in boundaryconformally invariant theories as well as for the classification ofAdS black holes preserving different fractions of supersymmetry. Journal volume: 16 Issue number: 05 Rights: World Scientific Publishing Company Publisher: World Scientific Publishing Company

Ferrarese, Laura; Merritt, David A Fundamental Relation between Supermassive Black Holes and Their Host Galaxies Journal The Astrophysical Journal Date: Aug 01 2000 Abstract: The masses of supermassive black holes correlate almost perfectly with the velocity dispersions of their host bulges. The relation is much tighter than the relation between Mbh and bulge luminosity, with a scatter no larger than expected on the basis of measurement error alone. Black hole masses recently estimated by Magorrian et al. lie systematically above the Mbh-relation defined by more accurate mass estimates, some by as much as 2 orders of magnitude. Journal volume: 539 Issue number: 1 Page number: L9-L12 Publisher: University of Chicago Press

Ferrari, Valeria; Pauri, Massimo; Piazza, Federico Quasinormal modes of charged, dilaton black holes Journal Physical Review Date: Mar 15 2001 Abstract: Abstract available at publisher web site Journal volume: 63 Issue number: 6 Page number: 064009-064009-13 Publisher: American Physical Society

Field, George Astrophysics Journal Reviews of Modern Physics Date: Mar 31 1999 Abstract: Abstract available at publisher web site Journal volume: 71 Issue number: 2 Page number: S33-S40 Publisher: American Physical Society

Firsova, N. E.; Sato, H. S -Matrices for Quantum Charged Massive Scalar Particles on Schwarzschild and Reissner-Nordström Black Holes Journal International Journal of Modern Physics Date: Feb 01 2000 Abstract: The authors study the scattering problems arising when considering the contribution of the topologically inequivalent configurations (TICs) of the massive complex scalar fields on Schwarzschild and Reissner-Nordström black holes to the Hawking radiation. The corresponding S-matrices are explored and presented in the form convenient to numerical computations. Journal volume: 09 Issue number: 01 Rights: World Scientific Publishing Company Publisher: World Scientific Publishing Company

Flambaum, V. V.; Berengut, J. C. Atom made from charged elementary black hole Journal Physical Review Date: Apr 15 2001 Abstract: Abstract available at publisher web site Journal volume: 63 Issue number: 8 Page number: 084010-084010-9 Publisher: American Physical Society

Fré, P.; Gorini, V.; and., P.Fré,V.Gorini,G.Magli; Moschella, U. Classical and Quantum Black Holes Journal Classical and Quantum Gravity Date: Jan 21 2000 Abstract: Black holes, first predicted two centuries ago, are a ubiquitous feature of modern gravitation theory. This volume reports the proceedings of a Spring School in 1998 which tried to give an almost encyclopaedic summary of black hole physics. P The Schwarzschild solution (1915) led on the one hand to an investigation of the gravitational collapse of stellar objects. Those with a mass less than about 1.4 solar masses can collapse to a stable state, a white dwarf, where gravity is balanced by the degeneracy pressure of free electrons. Heavier stars, whose mass is less than about three solar masses, can become neutron stars. Otherwise there is no stable configuration, and classical relativity predicts collapse to a gravitational singularity.

Frolov, V. P.; Fursaev, D. V. Mining energy from a black hole by strings Journal Physical Review Date: Jun 15 2001 Abstract: The authors discuss how cosmic strings can be used to mine energy from black holes. A string attached to the black hole gives rise to an additional channel for the energy release. It is demonstrated that when a string crosses the event horizon, its transverse degrees of freedom are thermally excited and thermal string perturbations propagate along the string to infinity. The internal metric induced on the 2D world sheet of the static string crossing the horizon describes a 2D black hole. Journal volume: 63 Issue number: 12 Page number: 124010-124010-7 Publisher: American Physical Society

Frolov, V. P.; Fursaev, D. V. Statistical mechanics on axially symmetric space-times with the Killing horizon and entropy of rotating black holes in induced gravity Journal Physical Review Date: Jan 15 2000 Abstract: Abstract available at publisher web site Journal volume: 61 Issue number: 2 Page number: 024007-024007-10 Publisher: American Physical Society

Frolov, V.; Fursaev, D. Black holes with polyhedral multi-string configurations Journal Classical and Quantum Gravity Date: Apr 21 2001 Abstract: P The authors find exact solutions of the Einstein equations which describe a black hole pierced by infinitely thin cosmic strings. The string segments enter the black hole along the radii and their positions coincide with the symmetry axes of a regular polyhedron. Each string produces an angle deficit proportional to its tension, while the metric outside the strings is locally a Schwarzschild one. There are three configurations corresponding to tetrahedra, octahedra and icosahedra where the number of string segments is 14, 26 and

62, respectively. There is also a 'double-pyramid' configuration where the number of string segments is not fixed. There can be two or three independent types of strings in one configuration. Tensions of strings belonging to the same type must be equal. Analogous polyhedral multi-string configurations can be combined with other spherically symmetric solutions of the Einstein equations. Journal volume: 18 Issue number: 8 Page number: 1535-1541 Publisher: Institute of Physics

Frolov, V.P.; Fursaev, D.V. Mechanism of the generation of black hole entropy in Sakharov's induced gravity Journal Physical Review, D (Particles Fields) Date: Aug 31 1997 Abstract: The mechanism of the generation of Bekenstein-Hawking entropy $S\{sup\ BH\}$ of a black hole in the Sakharov's induced gravity is proposed. It is suggested that the "physical" degrees of freedom, which explain the entropy $S\{sup\ BH\}$, form only a finite subset of the standard Rindler-like modes defined outside the black hole horizon. Journal volume: 56 Issue number: 4 Page number: 2212-2225 Subjects: Black Holes/entropy; Quantum Gravity/black holes; Entropy; Ultraviolet Divergences; Fluctuations; Hamiltonians; Space-Time; Probability; Bifurcation

Frolov, V.P.; Fursaev, D.V. Thermal fields, entropy and black holes Journal Classical and Quantum Gravity Date: Aug 01 1998 Abstract: In this review the authors describe the statistical mechanics of quantum systems in the presence of a Killing horizon and compare statistical-mechanical and 1-loop contributions to black-hole entropy. The study of these questions was motivated by attempts to explain the entropy of black holes as a statistical-mechanical entropy of quantum fields propagating near the black-hole horizon. The authors provide an introduction to this field of research and review its results. In particular, the authors discuss the relation between the statistical-mechanical entropy of quantum fields and the Bekenstein-Hawking entropy in the standard scheme with renormalization of gravitational coupling constants and in the theories of induced gravity. Journal volume: 15 Issue number: 8 Page number: 2041-2074 Publisher: Institute of Physics

Fryer, C.L.; Colgate, S.A.; Pinto, P.A. Iron Opacity and the Pulsar of SN 1987A Journal Astrophysical Journal Date: Feb 28 1999 Abstract: Neutron stars formed in Type II supernovae are likely to be initially obscured by late-time fallback. Although much of the late-time fallback is quickly accreted via neutrino cooling, some material remains on the neutron star, forming an atmosphere that slowly accretes through photon emission. In this paper, the authors derive structure equations of the fallback atmosphere and present results of one-dimensional simulations of that fallback. Journal volume: 511 Issue number: 2 Page number: 885-895 Subjects: Neutron Stars; Supernova Remnants; Pulsars; Star Evolution; Star Accretion; Stellar Atmospheres; Stellar Winds

Fryer, Chris L.; Woosley, S. E.; Hartmann, Dieter H. Formation Rates of Black Hole Accretion Disk Gamma-Ray Bursts Journal Astrophysical Journal Date: Nov 20 1999 Abstract: The cosmological origin of at least an appreciable fraction of classical gamma-ray bursts (GRBs) is now supported by redshift measurements for a half-dozen faint host galaxies. Still, the nature of the central engine (or engines) that provide the burst energy remains unclear. While many models have been proposed, those currently favored are all based upon the formation of and/or rapid accretion into stellar-mass black holes. Here the authors discuss a variety of such scenarios and estimate the probability of each.

G.W. Gibbons; Herdeiro, C.A.R. Supersymmetric rotating black holes and causality violation Journal Classical and Quantum Gravity Date: Nov 01 1999 Abstract: The geodesics of the rotating extreme black hole in five spacetime dimensions found by Breckenridge, Myers, Peet and Vafa are Liouville integrable and may be integrated by additively separating the Hamilton-Jacobi equation. This allows us to obtain the Stäckel-Killing tensor. The authors use these facts to give the maximal analytic extension of the spacetime and discuss some aspects of its causal structure. In particular, the authors exhibit a 'repulson'-like behaviour occurring when there are naked closed timelike curves.

Gaida, Ingo Quantum probes of repulsive singularities in supergravity Journal Classical and Quantum Gravity Date: Aug 01 1998 Abstract: Repulsive singularities ('repulsons') in extended supergravity theories are investigated. These repulsive singularities are related to attractive singularities ('black holes') in moduli space of extended supergravity vacua. In order to study these repulsive singularities, a scalar test-particle in the background of a repulson is investigated. It is shown, using a partial wave expansion, that the wavefunction of the scalar particle vanishes at the curvature singularity at the origin. In addition, the connection to higher-dimensional p-brane solutions including anti-branes is discussed. Journal volume: 15 Issue number: 8 Page number: 2261-2269 Publisher: Institute of Physics

Gaida, Ingo; Mahapatra, Swapna; Mohaupt, Thomas; Sabra, Wafic A. Black holes and flop transitions in M-theory on Calabi-Yau 3-folds Journal Classical and Quantum Gravity Date: Feb 01 1999 Abstract: The authors present five-dimensional extreme black hole solutions of M-theory compactified on Calabi-Yau 3-folds and study these solutions in the context of flop transitions in the extended Kähler cone. In particular, the authors consider a specific model and present black hole solutions, breaking half of $N = 2$ supersymmetry, in two regions of the extended Kähler cone, which are connected by a flop transition. The conditions necessary to match both solutions at the flop transition are analysed. Finally, the authors also discuss the conditions to obtain massless black holes at the flop transition. Journal volume: 16 Issue number: 2 Page number: 419-433 Publisher: Institute of Physics

Galloway, G. J.; Schleich, K.; Witt, D. M.; Woolgar, E. Topological censorship and higher genus black holes Journal Physical Review Date: Nov 15 1999 Abstract: Abstract available at publisher web site Journal volume: 60 Issue number: 10 Page number: 104039-104039-11 Publisher: American Physical Society

Gal'tsov, D.V.; Letelier, P.S. Interpolating black holes in dilaton - axion gravity Journal Classical and Quantum Gravity Date: Jan 01 1997 Abstract: A $2p + 5$ parametric family of black holes is constructed in dilaton - axion gravity with p vector fields using a holomorphic representation of U-duality in three dimensions. The metric of the non-extremal black holes has a Reissner - Nordström-type structure and generically possesses an internal horizon. However, in the extremal limit the generic solution exhibits a dilatonic-type null singularity. The solution thus interpolates between the well known Reissner - Nordström and charged dilatonic spacetimes. Journal volume: 14 Issue number: 1 Page number: L9-L14 Publisher: Institute of Physics

Gal'tsov, Dmitri V.; Lemos, José.P.S. No-go theorem for false vacuum black holes Journal Classical and Quantum Gravity Date: May 07 2001 Abstract: P The authors study the possibility of non-singular black hole solutions in the theory of general relativity coupled to a nonlinear scalar field with a positive potential possessing two minima: a 'false vacuum' with positive energy and a 'true vacuum' with zero energy. Assuming that the scalar field starts at the false vacuum at the origin and comes to the true vacuum at spatial infinity, the authors prove a no-go theorem by extending a no-hair theorem to the black hole interior: no smooth solutions exist which interpolate between the local de Sitter solution near the origin and the asymptotic Schwarzschild solution through a regular event horizon or several horizons. Journal volume: 18 Issue number: 9 Page number: 1715-1726. Institute of Physics

Gamboa, J.; Méndez, F. Scattering in three-dimensional extremal black holes Journal Classical and Quantum Gravity Date: Jan 21 2001 Abstract: P An approach to scattering theory in three-dimensional AdS spaces is proposed. Firstly, the authors consider the scattering of spinless relativistic particles by a three-dimensional extremal black hole and compute the absorption cross section $\sigma_{abs} = J_{abs}/J_\infty$; without using in and out states. Secondly, the authors posit a reciprocal space $\mathcal{H}$; where in and out states and the scattering amplitude are defined as in usual scattering theory. The authors show that both descriptions are equivalent and $\mathcal{H}$; could be considered as the space where the scattering processes in AdS should be defined. Journal volume: 18 Issue number: 2 Page number: 225-231 Publisher: Institute of Physics

Gao, Sijie; Wald, Robert M. "Physical process version" of the first law and the generalized second law for charged and rotating black holes Journal Physical Review Date: Oct 15 2001 Abstract: Abstract available at publisher web site Journal volume: 64 Issue number: 8 Page number: 084020-084020-14 Publisher: American Physical Society

Garay, L. J.; Anglin, J. R.; Cirac, J. I.; Zoller, P. Sonic Analog of Gravitational Black Holes in Bose-Einstein Condensates Journal Physical Review Letters Date: Nov 27 2000 Abstract: Abstract available at publisher web site Journal volume: 85 Issue number: 22 Page number: 4643-4647 Publisher: American Physical Society

Garay, L. J.; Anglin, J. R.; Cirac, J. I.; Zoller, P. Sonic black holes in dilute Bose-Einstein condensates Journal Physical Review A Date: Feb 28 2001 Abstract: Abstract available at publisher web site Journal volume: 63 Issue number: 2 Page number: 023611-023611-13 Publisher: American Physical Society

Garriga, J.; Vilenkin, A. In defense of the "tunneling" wave function of the universe Journal Physical Review, D (Particles Fields) Date: Aug 31 1997 Abstract: The tunneling approach to the wave function of the Universe has been recently criticized by Bousso and Hawking who claim that it predicts a catastrophic instability of de Sitter space with respect to pair production of black holes. The authors show that this claim is unfounded. Journal volume: 56 Issue number: 4 Page number: 2464-2468 Subjects: Cosmology; Universe; Tunneling; Instability; Pair Production; Quantum Gravity; WKB Approximation

Garriga, Jaume; Sasaki, Misao Brane-world creation and black holes Journal Physical Review Date: Aug 15 2000 Abstract: Abstract available at publisher web site Journal volume: 62 Issue number: 4 Page number: 043523-043523-8 Publisher: American Physical Society

Gauntlett, J.P.; Tada, T. Entropy of four-dimensional rotating BPS black hole dyons Journal Physical Review, D (Particles Fields) Date: Feb 28 1997 Abstract: The known BPS dyon black hole solutions of the N=4 heterotic string in four dimensions with a nonzero angular momentum all have naked singularities. The authors show that it is possible to modify a certain class of these solutions by the addition of massive Kaluza-Klein fields in such a way that the solutions decompactify near the core to five-dimensional black hole solutions with regular event horizons. Journal volume: 55 Issue number: 4 Page number: R1707-R1710 Subjects: Dyons; Entropy; Angular Momentum; Kaluza-Klein Theory

Gauntlett, Jerome P.; Kim, Nakwoo; Pakis, Stathis; Waldram, Daniel Membranes wrapped on holomorphic curves Journal Physical Review Date: Jan 15 2002 Abstract: Abstract available at publisher web site Journal volume: 65 Issue number: 2 Page number: 026003-026003-10 Publisher: American Physical Society

Gauntlett, Jerome P.; Myers, Robert C.; Townsend, Paul K. Black holes of D = 5 supergravity Journal Classical and Quantum Gravity Date: Jan 01 1999 Abstract: The authors discuss some general features of black holes of five-dimensional supergravity, such as the first law of black hole mechanics. The authors also discuss some special features of rotating supersymmetric black holes. In particular, the authors show that the horizon is a non-singular, and non-rotating, null hypersurface whose intersection with a Cauchy surface is a squashed 3-sphere. The authors find the Killing spinors of the near-horizon geometry and thereby determine the near-horizon isometry supergroup. Journal volume: 16 Issue number: 1 Page number: 1-21 Publisher: Institute of Physics

Ghosh, S. G.; Beesham, A. Higher dimensional inhomogeneous dust collapse and cosmic censorship Journal Physical Review Date: Dec 15 2001 Abstract: Abstract available at publisher web site Journal volume: 64 Issue number: 12 Page number: 124005-124005-5 Publisher: American Physical Society

Ghosh, S. G.; Dadhich, Naresh Naked singularities in higher dimensional Vaidya space-times Journal Physical Review Date: Aug 15 2001 Abstract: The authors investigate the end state of the gravitational collapse of a null fluid in higher-dimensional space-times. Both naked singularities and black holes are shown to be developing as the final outcome of the collapse. The naked singularity spectrum in a collapsing Vaidya region (4D) gets covered with the increase in dimensions and hence higher dimensions favor a black hole in comparison to a naked singularity. The cosmic censorship conjecture will be fully respected for a space of infinite dimension. Journal volume: 64 Issue number: 4 Page number: 047501-047501-4 Publisher: The American Physical Society

Gibbons, G.W. Collapsing shells and the isoperimetric inequality for black holes Journal Classical and Quantum Gravity Date: Oct 01 1997 Abstract: Recent results of Trudinger on isoperimetric inequalities for non-convex bodies are applied to the gravitational collapse of a lightlike shell of matter to form a black hole. Using some integral identities for co-dimension-two surfaces in Minkowski spacetime, the area A of the apparent horizon is shown to be bounded above in terms of the mass M by , which is consistent with the cosmic censorship hypothesis. The results hold in four spacetime dimensions and above. Journal volume: 14 Issue number: 10 Page number: 2905-2915 Publisher: Institute of Physics

Gibbons, G.W. Some comments on gravitational entropy and the inverse mean curvature flow Journal Classical and Quantum Gravity Date: Jun 01 1999 Abstract: The Geroch-Wald-Jang-Huisken-Ilmanen approach to the positive energy problem may be extended to give a negative lower bound for the mass of asymptotically anti-de Sitter spacetimes containing horizons with exotic topologies having ends or infinities of the form g ×, in terms of the cosmological constant. The authors also show how the method gives a lower bound for the mass of time-symmetric initial data sets for black holes with vectors and scalars in terms of the mass, |Z(Q,P)| of the double-extreme black hole with the same charges.

Giddings, Steven B.; Katz, Emanuel Effective theories and black hole production in warped compactifications Journal Journal of Mathematical Physics Date: Jul 01 2001 Abstract: The authors investigate aspects of the four-dimensional (4D) effective description of brane world scenarios based on warped compactification on anti-de Sitter space. The low-energy dynamics is described by visible matter gravitationally coupled to a "dark" conformal field theory. The authors give the linearized description of the 4D stress tensor corresponding to an arbitrary 5D matter distribution

Giddings, Steven B.; Lippert, Matthew Precursors, black holes, and a locality bound Journal Physical Review Date: Jan 15 2002 Abstract: Abstract available at publisher web site Journal volume: 65 Issue number: 2 Page number: 024006-024006-11 American Physical Society

Giddings, Steven B.; Ross, Simon F. D3-brane shells to black branes on the Coulomb branch Journal Physical Review Date: Jan 15 2000 Abstract: Abstract available at publisher web site tcal: N=4 SU(N) gauge theory which corresponds to a spherically symmetric shell of D3-branes. This point is of interest both because the spacetime region inside the shell is flat, and because this configuration gives a very simple example of the transition between D-branes in the perturbative string regime and the nonperturbative regime of black holes. The authors discuss how this geometry is described in the dual gauge theory, through its effect on the two-point functions and Wilson loops. In the calculation of the two-point function, the authors stress the importance of absorption by the branes. Journal volume: 61 Issue number: 2 Page number: 024036-024036-6 Publisher: American Physical Society

Glampedakis, Kostas; Andersson, Nils Scattering of scalar waves by rotating black holes Journal Classical and Quantum Gravity Date: May 21 2001 Abstract: The authors study the scattering of massless scalar waves by a Kerr black hole by letting plane monochromatic waves impinge on the black hole. The authors calculate the relevant scattering phase-shifts using the Prüfer phase-function method, which is computationally efficient and also reliable for high frequencies and/or large values of the angular multipole indices (l, m). The authors use the obtained phase-shifts and the partial-wave approach to determine differential cross sections and deflection functions. Results for off-axis scattering (waves incident along directions misaligned with the black hole's rotation axis) are obtained for the first time. Inspection of the off-axis deflection functions reveals the same scattering phenomena as in Schwarzschild scattering. Journal volume: 64 Issue number: 10 Page number: 104021-104021-20 Publisher: American Physical Society

Gnedin, Oleg Y. Million solar mass black holes at high redshift Journal Classical and Quantum Gravity Date: Oct 07 2001 Abstract: P The existence of quasars at redshift $z > 5$ indicates that supermassive black holes have been present since very early times. If they grew by accretion, the seeds of mass $>\text{rsim};105$ M must have formed at $z\sim9$. These seed black holes may result from the collapse and dissipation of primordial gas during the early stages of galaxy formation. The author discusses the effects of magnetic fields on the fragmentation of cold gas clouds embedded in a hot diffuse phase and a virialized dark matter halo.

Gomez, Roberto; Husa, Sascha; Winicour, Jeffrey Complete null data for a black hole collision Journal Physical Review Date: Jul 15 2001 Abstract: Abstract available at publisher web site Journal volume: 64 Issue number: 2 Page number: 024010-024010-20 Publisher: American Physical Society

Gonçalves, Sérgio M. C. V. Integrability of the minimal strain equations for the lapse and shift in 3+1 numerical relativity Journal Physical Review Date: Jul 15 2000 Abstract: Abstract available at publisher web site Journal volume: 62 Issue number: 2 Page number: 024009-024009-6 Publisher: American Physical Society

Gonçalves, Sérgio.M.C.V.; Moss, Ian G. Black hole formation from massive scalar fields Journal Classical and Quantum Gravity Date: Sep 01 1997 Abstract: It is shown that there exists a range of parameters in which gravitational collapse with a spherically symmetric massive scalar field can be treated as if it were a collapsing dust. This implies a criterion for the formation of black holes which is dependent on the size and mass of the initial field configuration and the mass of the scalar field. Journal volume: 14 Issue number: 9 Page number: 2607-2615 Publisher: Institute of Physics

Goncharov, Yu. P.; Firsova, N. E. Increase of the Hawking Radiation for Spinor Particles from Schwarzschild Black Holes by Dirac Monopoles Journal Modern Physics Letters A Date: Dec 01 2001 Abstract: An algorithm for numerical computation of the barrier transparencyfor the potentials surrounding Schwarzschild black holes is describedfor massless spinor particles. It is then applied to calculate all theconfigurations (including the contributions of twisted fieldconfigurations connected with Dirac monopoles) luminosity for theHawking radiation from a Schwarzschild black hole. It is found thatthe contribution due to monopoles can be of order 22% of all theconfigurations luminosity. Journal volume: 16 Issue number: 37 Rights: World Scientific Publishing Company Publisher: World Scientific Publishing Company

Gould, Andrew; Rix, Hans-Walter Binary Black Hole Mergers from Planet-like Migrations Journal The Astrophysical Journal Date: Mar 01 2000 Abstract: If supermassive black holes (BHs) are generically present in galaxy centers, and if galaxies are built up through hierarchical merging, BH binaries are at least temporary features of most galactic bulges. Observations suggest, however, that binary BHs are rare, pointing toward a binary lifetime far shorter than the Hubble time. The authors show that, almost regardless of the detailed mechanism, all stellar dynamical processes are too slow in reducing the orbital separation once orbital velocities in the binary exceed the virial velocity of the system.

Gour, Gilad Schwarzschild black hole as a grand canonical ensemble Journal Physical Review Date: Jan 15 2000 Abstract: Abstract available at publisher web site Journal volume: 61 Issue number: 2 Page number: 021501-021501-4 Publisher: American Physical Society

Govindarajan, T.R.; Suneeta, V. Quasi-normal modes of AdS black holes: a superpotential approach Journal Classical and Quantum Gravity Date: Jan 21 2001 Abstract: P A novel method, based on superpotentials is proposed for obtaining the quasi-normal modes of anti-de Sitter black holes. This is inspired by the case of the three-dimensional BTZ black hole, where the quasi-normal modes can be obtained exactly and are proportional to the surface gravity. Using this approach, the quasi-normal modes of the five-dimensional Schwarzschild anti-de Sitter black hole are computed numerically. The modes again seem to be proportional to the surface gravity for very small and very large black holes. They reflect the well known instability of small black holes in anti-de Sitter space. Journal volume: 18 Issue number: 2 Page number: 265-275 Publisher: Institute of Physics

Green, Anne M. Viability of primordial black holes as short period gamma-ray bursts Journal Physical Review Date: Jan 15 2002 Abstract: Abstract available at publisher web site Journal volume: 65 Issue number: 2 Page number: 027301-027301-4 Publisher: American Physical Society

Green, Anne M.; Malik, Karim A. Primordial black hole production due to preheating Journal Physical Review Date: Jul 15 2001 Abstract: Abstract available at publisher web site Journal volume: 64 Issue number: 2 Page number: 021301-021301-4 Publisher: American Physical Society

Gregory, James P.; Ross, Simon F. Looking for event horizons using UV-IR relations Journal Physical Review Date: May 15 2001 Abstract: Abstract available at publisher web site Journal volume: 63 Issue number: 10 Page number: 104023-104023-12 Publisher: American Physical Society

Griffiths, R. E.; Ptak, A.; Feigelson, E. D.; Garmire, G.; Townsley, L.; Brandt, W. N.; Sambruna, R.; Bregman, J. N. Hot Plasma and Black Hole Binaries in Starburst Galaxy M82 Journal Title: Science Magazine Date: Nov 17 2000 Abstract: Abstract available at publisher web site Journal volume: 290 Issue number: 5495 Page number: 1325-1328 DOE sponsorship Publisher: American Association for the Advancement of Science

Grumiller, D.; Kummer, W. Absolute conservation law for black holes Journal

Physical Review Date: Mar 15 2000 Abstract: Abstract available at publisher web site Journal volume: 61 Issue number: 6 Page number: 064006-064006-7 Publisher: American Physical Society

Gubser, S.S. Absorption of photons and fermions by black holes in four dimensions Journal Physical Review, Date: Dec 31 1997 Abstract: The absorption of photons and fermions into four-dimensional black holes is described by equations which in certain cases can be analyzed using dyadic index techniques. The resulting absorption cross sections for near-external black holes have a form at low energies suggestive of the effective string model. Journal volume: 5612 Page number: 7854-7868 Subject keyword: Absorption; Cross Sections; Fermions; Four-Dimensional Calculations; Photons; Spin; String Models; Supergravity; Supersymmetry; Angular Momentum; Bosons; Composite Models; Elementary Particles; Extended Particle Model; Field Theories; Massless Particles; Mathematical Models; Particle Models; Particle Properties; Quark Model; Sorption; Symmetry; Unified-Field Theories

Gubser, S.S. Can the effective string see higher partial waves? Journal Physical Review, D (Particles Fields) Date: Oct 31 1997 Abstract: The semiclassical cross sections for arbitrary partial waves of ordinary scalars to fall into certain five-dimensional black holes have a form that seems capable of explanation in terms of the effective string model. The kinematics of these processes is analyzed in detail on the effective string and is shown to reproduce the correct functional form of the semiclassical cross sections. Journal volume: 56 Issue number: 8 Page number: 4984-4993 Subjects: Semiclassical Approximation; Cross Sections; Scalars

Guido, D.; Longo, R.; Roberts, J. E.; Verch, R. Charged Sectors, Spin and Statisticsin Quantum Field Theory on Curved Spacetimes Journal Reviews in Mathematical Physics Date: Feb 01 2001 Abstract: The first part of this paper extends the Doplicher-Haag-Robertstheory of superselection sectors to quantum field theory on arbitraryglobally hyperbolic spacetimes. The statistics of a superselectionsector may be defined as in flat spacetime and each charge has aconjugate charge when the spacetime possesses non-compact Cauchysurfaces. In this case, the field net and the gauge group can beconstructed as in Minkowski spacetime. The second part of this paperderives spin-statistics theorems on spacetimes with appropriatesymmetries. Journal volume: 13 Issue number: 02 Rights: World Scientific Publishing Company Publisher: World Scientific Publishing Company

Gurtug, Ozay; Halilsoy, Mustafa Effect of sources on the inner horizon of black holes Journal Physical Review Date: Oct 15 2001 Abstract: Abstract available at publisher web site Journal volume: 64 Issue number: 8 Page number: 084023-084023-7 American Physical Society

Gutowski, J.; Papadopoulos, G. Moduli spaces for four- and five-dimensional black holes Journal Physical Review Date: Sep 15 2000 Abstract: Abstract available at publisher web site Journal volume: 62 Issue number: 6 Page number: 064023-064023-16 American Physical Society

H.W. Lee,Y.S. Myung, Jin Young Kim, Park, D.K. Quantum instability of two-dimensional charged black holes Journal Classical and Quantum Gravity Date: Mar 01 1997 Abstract: The authors study the quantum stability of a tachyon field in the two-dimensional (2D), charged black holes. In particular, the stress - energy tensor observed by a freely falling observer has new divergences such as and near the horizon , compared with conformal matter. It turns out that both the extremal and non-extremal black holes are unstable quantum mechanically. Journal volume: 14 Issue number: 3 Page number: L53-L58 Publisher: Institute of Physics

Haehnelt, Martin G.; Kauffmann, Guinevere The correlation between black hole mass and bulge velocity dispersion in hierarchical galaxy formation models Journal Monthly Notices of the Royal Astronomical Society Date: Nov 01 2000 Abstract: Recent work has demonstrated that there is a tight correlation between the mass of a black hole and the velocity dispersion of the bulge of its host galaxy. The authors show that the model of Kauffmann & Haehnelt, in which bulges and supermassive black holes both form during major mergers, produces a correlation between M bh and with a slope and scatter comparable to the observed relation.

Haiman, Zoltan; Abel, Tom; Rees, Martin J. The Radiative Feedback of the First Cosmological Objects Journal Astrophysical Journal Date: May 01 2000 Abstract: In hierarchical models of structure formation, an early cosmic UV background (UVB) is produced by the small (T_{vir} (less-or-similar sign)10^4 K) halos that collapse before re-ionization. The UVB at energies below 13.6 eV suppresses the formation of stars or black holes inside small halos by photo dissociating their only cooling agent, molecular H_2. The authors self-consistently compute the buildup of the early UVB in Press-Schechter models, coupled with H_2 photo dissociation both in the intergalactic medium (IGM) and inside virialized halos.

Halzen, F.; Zas, E. Neutrino fluxes from active galaxies: A model-independent estimate Journal Astrophysical Journal Date: Oct 31 1997 Abstract: There are tantalizing hints that jets, powered by supermassive black holes at the center of active galaxies, are true cosmic proton accelerators. They produce photons of TeV energy, possibly higher, and may be the enigmatic source of the highest energy cosmic rays. Journal volume: 488 Issue number: 2 Page number: 669-674 Subjects: Galactic Evolution; Cosmic Neutrinos

Hambli, N. Remarks on Dirichlet branes at angles Journal Physical Review, (Particles Fields) Date: Aug 31 1997 Abstract: This paper illustrates the derivation of the low-energy background field solutions of D2-branes and D4-branes intersecting at nontrivial angles by solving directly the bosonic equations of motion of II supergravity coupled to a dilaton and antisymmetric fields. The authors also argue for how a similar analysis can be performed for any similar Dp-branes oriented at angles. Journal volume: 56 Issue number: 4 Page number: 2369-2377 Subjects: Bosons; Supergravity; Entropy

Hansen, R. N.; Christensen, M.; Larsen, A. L. Comment on "Formation of primordial black holes by cosmic strings" Journal Physical Review Date: May 15 2000 Abstract: Abstract available at publisher web site Journal volume: 61 Issue number: 10 Page number: 108701-108701-2 Publisher: American Physical Society

Hansen, R. N.; Christensen, M.; Larsen, A. L. Cosmic String Loops Collapsing to Black Holes Journal International Journal of Modern Physics A Date: Nov 10 2000 Abstract: The authors examine the question of collapse of Turok's two-parameter family of cosmic strings. The authors first perform a classification of the strings according to the specific time(s) at which the minimal string size is reached during one period. The authors then obtain an exact analytical expression for the probability of collapse to black holes for the Turok strings. Our result has the same general behavior as previously obtained in the literature but the authors find, in addition, a numerical prefactor that changes the result by three orders of magnitude. Finally the authors show that our careful computation of the prefactor helps us to understand the discrepancy between previously obtained results and, in particular, that for "large" values of G, there may not even be a discrepancy. The authors also give a simple physical argument that can immediately rule out some of the previously obtained results. Journal volume: 15 Issue number: 28 Rights: World Scientific Publishing Company Publisher: World Scientific Publishing Company

Hartmann, Betti; Kleihaus, Burkhard; Kunz, Jutta Axially symmetric monopoles and black holes in Einstein-Yang-Mills-Higgs theory Journal Physical Review Date: Jan 15 2002 Abstract: Abstract available at publisher web site Journal volume: 65 Issue number: 2 Page number: 024027-024027-22 Publisher: American Physical Society

Hawking, S. W.; Reall, H. S. Charged and rotating AdS black holes and their CFT duals Journal Physical Review Date: Jan 15 2000 Abstract: Abstract available at publisher web site Journal volume: 61 Issue number: 2 Page number: 024014-024014-10 Publisher: American Physical Society

Hawking, S.W. Stability of AdS and phase transitions Journal Classical and Quantum Gravity Date: Mar 07 2000 Abstract: Black holes are often thought of as completely dead classically. That is, they absorb, but do not give out, radiation and energy. In this paper, first delivered at the Strings '99 Conference, in Potsdam, Germany, Professor Hawking explains that this is not necessarily the case. Journal volume: 17 Issue number: 5 Page number: 1093-1099 Publisher: Institute of Physics

Hayward, Sean A. An extreme critical spacetime: echoing and black-hole perturbations Journal Classical and Quantum Gravity Date: Oct 07 2000 Abstract: P A homothetic, static, spherically symmetric solution to the massless Einstein-Klein-Gordon equations is described. There is a curvature singularity which is central, null, bifurcate, massless and marginally trapped. The spacetime is therefore extreme in the sense of lying at the threshold between black holes and naked singularities, just avoiding both. A linear perturbation analysis reveals two types of dominant mode.

Hayward, Sean A. Gravitational waves from quasispherical black holes Journal Physical Review Date: May 15 2000 Abstract: Abstract available at publisher web site Journal volume: 61 Issue number: 10 Page number: 101503-101503-4 Publisher: American Physical Society

Hayward, Sean A. Gravitational waves, black holes and cosmic strings in cylindrical symmetry Journal Classical and Quantum Gravity Date: Apr 21 2000 Abstract: Gravitational waves in cylindrically symmetric Einstein gravity are described by an effective energy tensor with the same form as that of a massless Klein-Gordon field, in terms of a gravitational potential generalizing the Newtonian potential. Energy-momentum vectors for the gravitational waves and matter are defined with respect to a canonical flow of time. The combined energy-momentum is covariantly conserved, the corresponding charge being the modified Thorne energy. Energy conservation is formulated as the first law expressing the gradient of the energy as work and energy-supply terms, including the energy flux of the gravitational waves.

Hayward, Sean A. Gravitational-wave dynamics and black-hole dynamics: second quasi-spherical approximation Journal Classical and Quantum Gravity Date: Dec 21 2001 Abstract: Gravitational radiation with roughly spherical wavefronts, produced by roughly spherical black holes or other astrophysical objects, is described by an approximation scheme. The first quasi-spherical approximation, describing radiation propagation on a background, is generalized to include additional non-linear effects, due to the radiation itself. The gravitational radiation is locally defined and admits an energy tensor, satisfying all standard local energy conditions and entering the truncated Einstein equations as an effective energy tensor. This second quasi-spherical approximation thereby includes gravitational radiation reaction, such as the back-reaction on the black hole. With respect to a canonical flow of time, the combined energy-momentum of the matter and gravitational radiation is covariantly conserved.

Hayward, Sean A. Unified first law of black-hole dynamics and relativistic thermodynamics Journal Classical and Quantum Gravity Date: Oct 01 1998 Abstract: A unified first law of black-hole dynamics and relativistic thermodynamics is derived in spherically symmetric general relativity. This equation expresses the gradient of the active gravitational energy E according to the Einstein equation, divided into energy-supply and work terms. Projecting the equation along the flow of thermodynamic matter and along the trapping horizon of a black hole yield, respectively, first laws of relativistic thermodynamics and black-hole dynamics. In the black-hole case, this first law has the same form as the first law of black-hole statics, with static perturbations replaced by the derivative along the horizon.

Heckler, A.F. Calculation of the Emergent Spectrum and Observation of Primordial Black Holes Journal Physical Review Letters Date: May 31 1997 Abstract: The authors calculate the emergent spectrum of microscopic black holes, taking into account the proposition that a quark and gluon photosphere forms around the black hole. The authors find that the limit on the average universal density of black holes is not significantly affected by the photosphere. Journal volume: 78 Issue number: 18 Page number: 3430-3433 Subjects:Spectra; Quarks; Gluons; Gamma Radiation; Galaxies

Helfer, Adam D. Redshifts near black holes Journal Classical and Quantum Gravity Date: Dec 21 2001 Abstract: PA simple ordinary differential equation is derived governing the redshifts of wavefronts propagating through a non-stationary spherically symmetric space-time. Approach to an event horizon corresponds to approach to a fixed point; in general, the phase portrait of the equation illuminates the qualitative features of the geometry. In particular, the asymptotics of the redshift as the horizon is approached, a critical ingredient of Hawking's prediction of radiation from black holes, are easily brought out.

Hemming, Samuli; Keski-Vakkuri, Esko Hawking radiation from AdS black holes Journal Physical Review Date: Aug 15 2001 Abstract: The authors investigate Hawking radiation from black holes in (d+1)-dimensional anti--de Sitter space. The authors focus on s waves, make use of the geometrical optics approximation, and follow three approaches to analyze the radiation. First, the authors compute a Bogoliubov transformation between Kruskal and asymptotic coordinates and compare the different vacua. Second, following a method due to Kraus, Parikh, and Wilczek, the authors view Hawking radiation as a tunneling process across the horizon and compute the tunneling probability.

Higuchi, Atsushi Low-frequency scalar absorption cross sections for stationary black holes Journal Classical and Quantum Gravity Date: Oct 21 2001 Abstract: P The authors discuss the absorption cross section for the minimally-coupled massless scalar field into a stationary and circularly symmetric black hole with nonzero angular velocity in four or higher dimensions. In particular, the authors show that it equals the horizon area in the zero-frequency limit provided that the solution of the scalar field equation with an incident monochromatic plane wave converges pointwise to a smooth time-independent solution outside the black hole and on the future horizon, with the error term being at most linear in the frequency. The authors also show that this equality holds for static black holes which are not necessarily spherically symmetric. The zero-frequency scattering cross section is found to vanish in both cases. Journal volume: 18 Issue number: 20 Page number: L139-L144 Publisher: Institute of Physics

Hirayama, Takayuki; Kang, Gungwon Stable black strings in anti-de Sitter space Journal Physical Review Date: Sep 15 2001 Abstract: Abstract available at publisher web site Journal volume: 64 Issue number: 6 Page number: 064010-064010-8 Publisher: American Physical Society

Hjelmeland, S.E. Three-dimensional black hole from a stringy anti–de Sitter background Journal Physical Review, D (Particles Fields) Date: Feb 28 1997 Abstract: A new black hole solution in 2+1 dimensions is found by taking cosmic strings as part of the vacuum structure of the anti–de Sitter space-time. The solution has a structure that in many ways is similar to that of the Reissner-Nordström solution. With a vanishing cosmological constant, a space-time with a black hole of infinite extension appears with the inner horizon playing the role of a cosmological event horizon. Journal volume: 55 Issue number: 4 Page number: 2099-2104 Subjects: Differential Geometry;Cosmology; Vacuum States; Space-Time; Quantum Gravity; General Relativity Theory

Ho, Jeongwon Proof of the generalized second law for two-dimensional black holes Journal Physical Review Date: Sep 15 2001 Abstract: Abstract available at publisher web site Journal volume: 64 Issue number: 6 Page number: 064019-064019-6 American Physical Society

Hod, S; Piran, T. Critical behavior and universality in gravitational collapse of a charged scalar field Journal Physical Review, D (Particles Fields) Date: Mar 31 1997 Abstract: The authors summarize results from a study of spherically symmetric collapse of a *charged* (complex) massless scalar field. The authors present an analytic argument which conjectures the generalization of the mass-scaling relation and echoing phenomena, originally discovered by Choptuik, for the *charged* case. Journal volume: 55 Issue number: 6 Page number: 3485-3496 Subjects: Scalar Fields/Gravitational Collapse; Gravitational Charged Particles

Hod, Shahar Evidence for a null entropy of extremal black holes Journal Physical Review Date: Apr 15 2000 Abstract: Abstract available at publisher web site Journal volume: 61 Issue number: 8 Page number: 084018-084018-4 Publisher: American Physical Society

Holst, Sören; Peldán, Peter Black holes and causal structure in anti-de Sitter isometric spacetimes Journal Classical and Quantum Gravity Date: Dec 01 1997 Abstract: The observation that the $(2 + 1)$-dimensional BTZ black hole can be obtained as a quotient space of anti-de Sitter space leads one to ask what causal behaviour other such quotient spaces can display. In this paper the authors answer this question in $2 + 1$ and $3 + 1$ dimensions when the identification group has one generator. Among other things the authors find that there does not exist any $3 + 1$ generalization of the rotating BTZ hole. However, the non-rotating generalization exists and exhibits some unexpected properties; for example, it turns out to be non-static and to possess a non-trivial apparent horizon. Journal volume: 14 Issue number: 12 Page number: 3433-3452 Publisher: Institute of Physics

Hong, Soon-Tae; Kim, Won Tae; Kim, Yong-Wan; Park, Young-Jai Global embeddings of scalar-tensor theories in 2+1 dimensions Journal Physical Review Date: Sep 15 2000 Abstract: Abstract available at publisher web site Journal volume: 62 Issue number: 6 Page number: 064021-064021-9 Publisher: American Physical Society

Hong, Soon-Tae; Kim, Won Tae; Oh, John J.; Park, Young-Jai Higher dimensional flat embeddings of black strings in Journal Physical Review Date: Jun 15 2001 Abstract: The authors obtain (3+1)- and (3+2)-dimensional global flat embeddings of 2+1 uncharged and charged black strings, respectively. In particular, the charged black string, which is the dual solution of the Banados-Teitelboim-Zanelli black holes, is shown to be embedded in the same global embedding Minkowski space structure as that of the 2+1 charged de Sitter black hole solution. Journal volume: 63 Issue number: 12 Page number: 127502-127502-4 Publisher: The American Physical Society

Hong, Soon-Tae; Kim, Won Tae; Oh, John J.; Park, Young-Jai Higher dimensional flat embeddings of black strings in 2+1 dimensions Journal Physical Review Date: Jun 15 2001 Abstract: Abstract available at publisher web site Journal volume: 63 Issue number: 12 Page number: 127502-127502-4 Publisher: American Physical Society

Hong, Soon-Tae; Kim, Yong-Wan; Park, Young-Jai Higher dimensional flat embeddings of (2+1)-dimensional black holes Journal Physical Review Date: Jul 15 2000 Abstract: Abstract available at publisher web site Journal volume: 62 Issue number: 2 Page number: 024024-024024-6 American Physical Society

Horowitz, G.T.; Martinec, E.J. Comments on black holes in matrix theory Journal Physical Review, D (Particles Fields) Date: Apr 30 1998 Abstract: The recent suggestion that the entropy of Schwarzschild black holes can be computed in matrix theory using near-extremal D-brane thermodynamics is examined. It is found that the regime in which this approach is valid actually describes black strings stretched across the longitudinal direction, near the transition where black strings become unstable to the formation of black holes. Journal volume: 57 Issue number: 8 Page number: 4935-4941 Subjects: Supersymmetry; Schwarzschild Metric; Matrix Elements; Thermodynamics; Yang-Mills Theory

Horowitz, G.T.; Polchinski, J. Correspondence principle for black holes and strings Journal Physical Review, D (Particles Fields) Date: May 31 1997 Abstract: For most black holes in string theory, the Schwarzschild radius in string units decreases as the string coupling is reduced. The authors formulate a correspondence principle, which states that (i) when the size of the horizon drops below the size of a string, the typical black hole state becomes a typical state of strings and D-branes with the same charges, and (ii) the mass does not change abruptly during the transition. This provides a statistical interpretation of black hole entropy. Journal volume: 55 Issue number: 10 Page number: 6189-6197 Subjects: Coupling; Schwarzschild Radius; Entropy

Horowitz, G.T.; Ross, S.F. Naked black holes Journal Physical Review, Date: Aug 31 1997 Abstract: It is shown that there are large static black holes for which all curvature invariants are small near the event horizon, yet any object which falls in experiences enormous tidal forces *outside* the horizon. These black holes are charged and near extremality, and exist in a wide class of theories including string theory. Journal volume: 564 Page number: 2180-2187 Subjects: Cosmology; Space-Time; Supergravity; Composite Models; Quark Model; Unified-Field Theories

Horowitz, G.T; Yang, H. Black strings and classical hair Journal Physical Review, D (Particles Fields) Date: Jun 30 1997 Abstract: The authors examine the geometry near the event horizon of a family of black string solutions with traveling waves. It has previously been shown that the metric is continuous there. Contrary to expectations, the authors find that the geometry is not smooth, and the horizon becomes singular whenever a wave is present. Journal volume: 55 Issue number: 12 Page number: 7618-7624 Subjects: Cosmology; Geometry; Travelling Waves

Horowitz, Gary T. Comments on black holes in string theory Journal Classical and Quantum Gravity Date: Mar 07 2000 Abstract: A very brief review is given of some of the developments leading to our current understanding of black holes in string theory. This is followed by a discussion of two possible misconceptions in this subject - one involving the stability of small black holes and the other involving scale radius duality. Finally, I describe some recent results concerning quasinormal modes of black holes in anti-de Sitter spacetime, and their implications for strongly coupled conformal field theories (in various dimensions). Journal volume: 17 Issue number: 5 Page number: 1107-1116 Publisher: Institute of Physics

Horowitz, Gary T.; Hubeny, Veronika E. Quasinormal modes of AdS black holes and the approach to thermal equilibrium Journal Physical Review Date: Jul 15 2000 Abstract: Abstract available at publisher web site Journal volume: 62 Issue number: 2 Page number: 024027-024027-11 Publisher: American Physical Society

Horowitz, Gary T.; Maeda, Kengo Fate of the Black String Instability Journal Physical Review Letters Date: Sep 24 2001 Abstract: Abstract available at publisher web site Journal volume: 87 Issue number: 13 Page number: 131301-131301-4 Publisher: American Physical Society

Horowitz, Gary T.; Teukolsky, Saul A. Black holes Journal Reviews of Modern Physics Date: Mar 31 1999 Abstract: Abstract available at publisher web site Journal volume: 71 Issue number: 2 Page number: S180-S186 American Physical Society

Huang, Wung-Hong Microcanonical statistics of black holes and the bootstrap condition Journal Physical Review Date: Aug 15 2000 Abstract: Abstract available at publisher web site Journal volume: 62 Issue number: 4 Page number: 043002-043002-8 Publisher: American Physical Society

Hübner, Peter How to avoid artificial boundaries in the numerical calculation of black hole spacetimes Journal Classical and Quantum Gravity Date: Jul 01 1999 Abstract: This is the first of a series of papers describing a numerical implementation of the conformally rescaled Einstein equation, an implementation designed to calculate asymptotically flat spacetimes, especially spacetimes containing black holes. P The authors derive the new first-order time evolution equations to be used in the scheme. These time evolution equations can either be written in symmetric hyperbolic or in flux-conservative form. Since the conformally rescaled Einstein equation, also called the conformal field equations, formally allow us to place the grid boundaries outside the physical spacetime, the authors can modify the equations near the grid boundaries and obtain a consistent and stable discretization.

Hübner, Peter Numerical calculation of conformally smooth hyperboloidal data Journal Classical and Quantum Gravity Date: Apr 21 2001 Abstract: P This is the third paper in a series describing a numerical implementation of the conformal Einstein equation. This paper describes a scheme to calculate (three-)dimensional data for the conformal field equations from a set of free functions. The actual implementation depends on the topology of the spacetime. The authors discuss the implementation and exemplary calculations for data leading to spacetimes with one spherical null infinity (asymptotically Minkowski) and for data leading to spacetimes with two toroidal null infinities (asymptotically A3). The authors also outline the (technical) modifications of the implementation needed to calculate data for spacetimes with two and more spherical null infinities (asymptotically Schwarzschild and asymptotically multiple black holes). Journal volume: 18 Issue number: 8 Page number: 1421-1440 Publisher: Institute of Physics

Hughes, Scott A. Computing radiation from Kerr black holes: Generalization of the Sasaki-Nakamura equation Journal Physical Review Date: Aug 15 2000 Abstract: Abstract available at publisher web site Journal volume: 62 Issue number: 4 Page number: 044029-044029-8 Publisher: American Physical Society

Hughes, Scott A. Erratum: Evolution of circular, nonequatorial orbits of Kerr black holes due to gravitational-wave emission [Phys. Rev. D 61, 084004 (2000)] Journal Physical Review Date: Feb 15 2001 Abstract: Abstract available at publisher web site Journal volume: 63 Issue number: 4 Page number: 049902-049902-1 Publisher: American Physical Society

Hughes, Scott A. Evolution of circular, nonequatorial orbits of Kerr black holes due to gravitational-wave emission Journal Physical Review Date: Apr 15 2000 Abstract: Abstract available at publisher web site Journal volume: 61 Issue number: 8 Page number: 084004-084004-28 Publisher: American Physical Society

Hughes, Scott A. Evolution of circular, nonequatorial orbits of Kerr black holes due to gravitational-wave emission. II. Inspiral trajectories and gravitational waveforms Journal Physical Review Date: Sep 15 2001 Abstract: Abstract available at publisher web site Journal volume: 64 Issue number: 6 Page number: 064004-064004-15 Publisher: American Physical Society

Hughes, Scott A. Nearly horizon skimming orbits of Kerr black holes Journal Physical Review Date: Mar 15 2001 Abstract: Abstract available at publisher web site Journal volume: 63 Issue number: 6 Page number: 064016-064016-9 Publisher: American Physical Society

Hughes, Scott A. Untangling the merger history of massive black holes with LISA Journal Monthly Notices of the Royal Astronomical Society Date: Jan 31 2002 Abstract: Abstract available at Blackwell Science. Journal volume: 331 Issue number: 3 Page number: 805-816 DOE sponsorship Subjects: black hole physics; gravitation; gravitational waves; cosmology: miscellaneous Publisher: Blackwell Science, Ltd

Husa, Sascha; Winicour, Jeffrey Asymmetric merger of black holes Journal Physical Review Date: Oct 15 1999 Abstract: Abstract available at publisher web site Journal volume: 60 Issue number: 8 Page number: 084019-084019-13 Publisher: American Physical Society

Iizuka, Norihiro; Kabat, Daniel; Lifschytz, Gilad; Lowe, David A. Probing black holes in nonperturbative gauge theory Journal Physical Review Date: Jan 15 2002 Abstract: Abstract available at publisher web site Journal volume: 65 Issue number: 2 Page number: 024012-024012-14 Publisher: American Physical Society

Iwasawa, K.; Fabian, A. C.; Almaini, O.; Lira, P.; Lawrence, A.; Hayashida, K.; Inoue, H. X-ray absorption and rapid variability of the dwarf Seyfert nucleus of NGC 4395 Journal Monthly Notices of the Royal Astronomical Society Date: Nov 01 2000 Abstract: The authors report the detection of an absorbed central X-ray source and its strong, rapid, variability in NGC 4395, the least luminous Seyfert nucleus known. The X-ray source exhibits a number of flares with factors of 3-4 flux changes during a half-day ASCA observation. The shortest doubling time observed is about 100 s. Such X-ray variability is in contrast to the behaviour of other low-luminosity active galaxies and resembles that of higher luminosity Seyfert 1 galaxies. It provides further support for an accreting black hole model rather than an extreme stellar process in accounting for the nuclear activity of NGC 4395.

Jackson, Mark G. Two black hole holography, lensing, and intensity Journal Physical Review Date: Aug 15 2001 Abstract: The authors numerically verify the analysis of the 'expanding horizon' theory of Susskind in relation to the 't Hooft holographic conjecture. By using a numerical simulation to work out the holographic image formed by two black holes upon a screen very far away, it is seen that it is impossible for a horizon to hide behind another. The authors also compute the holographic intensity distribution of such an arrangement. Journal volume: 64 Issue number: 4 Page number: 044020-044020-7 Publisher: The American Physical Society

Jacobson, Ted Primordial Black Hole Evolution in Tensor-Scalar Cosmology Journal Physical Review Letters Date: Oct 04 1999 Abstract: Abstract available at publisher web site Journal volume: 83 Issue number: 14 Page number: 2699-2702 Publisher: American Physical Society

Jalali, M. A.; Rafiee, A. R. Integrable models of galactic discs with double nuclei Journal Monthly Notices of the Royal Astronomical Society Date: Jan 01 2001 Abstract: The authors introduce a new class of 2D mass models, whose potentials are of Stäckel form in elliptic coordinates. Our model galaxies have two separate strong cusps that form double nuclei. The potential and surface density distributions are locally axisymmetric near the nuclei and become highly non-axisymmetric outside the nucleus. The surface density diverges toward the cuspy nuclei with the law &formmu1; Our model is sustained by four general types of regular orbits: butterfly, nucleophilic banana, horseshoe and aligned loop orbits.

Jalali, Mir Abbas; de Zeeuw, P. Tim Self-consistent axisymmetric Sridhar-Touma models Journal Monthly Notices of the Royal Astronomical Society Date: Dec 01 2001 Abstract: The authors construct phase-space distribution functions for the oblate, cuspy mass models of Sridhar & Touma (ST), which may contain a central point mass (black hole) and have potentials of Stäckel form in parabolic coordinates. The density in the ST models is proportional to a power $r^{-\gamma}$ of the radius, with &formmu1;. The authors derive distribution functions $f(E, L_z)$ for the scale-free ST models (no black hole) using a power series of the energy E and the component L_z of the angular momentum parallel to the symmetry axis.

Jaranowski, Piotr; Schäfer, Gerhard Bare masses in time-symmetric initial-value solutions for two black holes Journal Physical Review Date: Mar 15 2000 Abstract: Abstract available at publisher web site Journal volume: 61 Issue number: 6 Page number: 064008-064008-3 Publisher: American Physical Society

Jaranowski, Piotr; Schäfer, Gerhard Binary black-hole problem at the third post-Newtonian approximation in the orbital motion: Static part Journal Physical Review Date: Dec 15 1999 Abstract: Abstract available at publisher web site Journal volume: 60 Issue number: 12 Page number: 124003-124003-7 Publisher: American Physical Society

Jedamzik, K. Primordial black hole formation during the QCD epoch Journal Physical Review, D (Particles Fields) Date: May 31 1997 Abstract: The authors consider the formation of horizon-size primordial black holes (PBH's) from pre-existing density fluctuations during cosmic phase transitions. It is pointed out that the formation of PBH's should be particularly efficient during the QCD epoch due to a substantial reduction of pressure forces during adiabatic collapse, or equivalently, a significant decrease in the effective speed of sound during the color-confinement transition. Journal volume: 55 Issue number: 10 Page number: R5871-R5875 Subjects: Bag Model; Cosmology; Nonluminous Matter; Density; Fluctuations;Confinement; Perturbation Theory; Spectra; Mass; Nucleosynthesis;

Jenet, F. A.; Prince, T. A. Detection of variable frequency signals using a fast chirp transform Journal Physical Review Date: Dec 15 2000 Abstract: Abstract available at publisher web site Journal volume: 62 Issue number: 12 Page number: 122001-122001-10 Publisher: American Physical Society

Jhingan, S.; Dadhich, N.; Joshi, P. S. Gravitational collapse in a constant potential bath Journal Physical Review Date: Feb 15 2001 Abstract: Abstract available at publisher web site Journal volume: 63 Issue number: 4 Page number: 044010-044010-6 Publisher: American Physical Society

Jhingan, S.; Magli, G. Black holes versus naked singularities formation in collapsing Einstein clusters Journal Physical Review Date: Jun 15 2000 Abstract: Abstract available at publisher web site Journal volume: 61 Issue number: 12 Page number: 124006-124006-7 Publisher: American Physical Society

Jing, Jiliang; Yan, Mu-Lin Effect of spin on the quantum entropy of black holes Journal Physical Review Date: Apr 15 2001 Abstract: Abstract available at publisher web site Journal volume: 63 Issue number: 8 Page number: 084028-084028-10 Publisher: American Physical Society

Jing, Jiliang; Yan, Mu-Lin Entropies of rotating charged black holes from conformal field theory at Killing horizons Journal Physical Review Date: Nov 15 2000 Abstract: Abstract available at publisher web site Journal volume: 62 Issue number: 10 Page number: 104013-104013-6 Publisher: American Physical Society

Jing, Jiliang; Yan, Mu-Lin Quantum entropy of a nonextreme stationary axisymmetric black hole due to a minimally coupled quantum scalar field Journal Physical Review Date: Oct 15 1999 Abstract: Abstract available at publisher web site Journal volume: 60 Issue number: 8 Page number: 084015-084015-7 Publisher: American Physical Society

Johnson, Clifford V.; Myers, Robert C. The enhance on, black holes, and the second law Journal Physical Review Date: Nov 15 2001 Abstract: Abstract available at publisher web site Journal volume: 64 Issue number: 10 Page number: 106002-106002-11 Publisher: American Physical Society

Joshi, P.S.; Dwivedi, I.H. Initial data and the end state of spherically symmetric gravitational collapse Journal Classical and Quantum Gravity Date: Jan 01 1999 Abstract: Generalizing earlier results on the initial data and the final fate of dust collapse, the authors study here the relevance of the initial state of a spherically symmetric matter cloud towards determining its end state in the course of a continuing gravitational collapse. It is shown that given an arbitrary regular distribution of matter at the initial epoch, there always exists an evolution from this initial data which would result either in a black hole or a naked singularity, depending on the allowed choice of free functions available in the solution. It follows that given any initial density and pressure profiles for the cloud, there is a non-zero measure set of configurations leading either to black holes or naked singularities, subject to the usual energy conditions ensuring the positivity of energy density. The authors also characterize here wide new families of black hole solutions resulting from spherically symmetric collapse without requiring the cosmic censorship assumption. Journal volume: 16 Issue number: 1 Page number: 41-59 Publisher: Institute of Physics

Joshi, Pankaj S.; Dadhich, Naresh K.; Maartens, Roy Gamma-Ray Bursts as the Birth-Cries of Black Holes Journal Modern Physics Letters A Date: May 20 2000 Abstract: The origin of cosmic gamma-ray bursts remains one of the most intriguing puzzles in astronomy. The authors suggest that purely general relativistic effects in the collapse of massive stars could account for these bursts. The late formation of closed trapped surfaces can occur naturally, allowing the escape of huge energy from curvature-generated fireballs, before these are hidden within a black hole. Journal volume: 15 Issue number: 15 Rights: World Scientific Publishing Company Publisher: World Scientific Publishing Company

K., G. Diffractive/refractive optics for high energy astronomy Journal Astronomy and Astrophysics Date: May 22 2001 Abstract: Diffractive optics components such as Fresnel Zone Plates and their derivatives potentially form the basis for telescope systems for X-ray and gamma-ray astronomy with high sensitivity and super bangular resolution. The main problem is that systems with convenient design parameters involve very long focal lengths. The design considerations and performance of telescopes using a simple Phase Fresnel Lens on one spacecraft and a detector assembly on another are considered.

Kallosh, R. Bound states of branes with minimal energy Journal Physical Review,

D (Particles Fields) Date: Mar 31 1997 Abstract: It is pointed out that the energy of the bound states of D-branes and strings is determined by the central charge of the space-time supersymmetry. The universality which is seen at the black hole horizon appears also on the D-brane side: the total energy of the bound states of a given number of branes has a minimum when considered as a function of the independent parameters (moduli). This provides new evidence that the near-horizon space-time geometry of the dilaton black holes can be represented by the bound states of branes. Journal volume: 55 Issue number: 6 Page number: R3241-R3245 Subjects: Bound State/energy; String Models/bound state; Energy; Space-Time; Supersymmetry; Geometry; Black Holes; Solitons

Kallosh, R. Bound states of branes with minimal energy Journal Physical Review, Date: Mar 31 1997 Abstract: It is pointed out that the energy of the bound states of D-branes and strings is determined by the central charge of the space-time supersymmetry. The universality which is seen at the black hole horizon appears also on the D- brane side: the total energy of the bound states of a given number of branes has a minimum when considered as a function of the independent parameters (moduli). This provides new evidence that the near-horizon space-time geometry of the dilaton black holes can be represented by the bound states of branes. Journal volume: 556 Page number: R3241-R3245 Subjects: Bound State_ Energy; String Models-- Bound State; Black Holes; Geometry; Solitons; Space-Time; Supersymmetry; Composite Models; Extended Particle Model; Mathematical Models; Mathematics; Particle Models; Quark Model; Quasi Particles; Symmetry

Kallosh, R.; Rajaraman, A.; Kai Wong, W. Supersymmetric rotating black holes and attractors Journal Physical Review, D (Particles Fields) Date: Mar 31 1997 Abstract: Five-dimensional stringy rotating black holes are embedded into N=2 supergravity interacting with one vector multiplet. The existence of an unbroken supersymmetry of the rotating solution is proved directly by solving the Killing spinor equations. The asymptotic enhancement of supersymmetry near the horizon in the presence of rotation is established via the calculation of the supercurvature. Journal volume: 55 Issue number: 6 Page number: R3246-R3249 Subjects: Black Holes/supergravity; Supergravity; Many-Dimensional Calculations; Rotation; String Models; Vectors; Multiplets; Supersymmetry; Attractors

Kallosh, R.; Rajaraman, A.; Kai Wong, W. Supersymmetric rotating black holes and attractors Journal Physical Review, D (Particles Fields) Date: Mar 31 1997 Abstract: Five-dimensional stringy rotating black holes are embedded into N=2 supergravity interacting with one vector multiplet. The existence of an unbroken supersymmetry of the rotating solution is proved directly by solving the Killing spinor equations. The asymptotic enhancement of supersymmetry near the horizon in the presence of rotation is established via the calculation of the supercurvature. Journal volume: 55 Issue number: 6 Page number: R3246-R3249 Subjects: Black Holes/supergravity; Supergravity; Many-Dimensional Calculations; Rotation; String Models; Vectors; Multiplets; Supersymmetry; Attractors

Kallosh, R; Linde, A. Black hole superpartners and fixed scalars Journal Physical Review, D (Particles Fields) Date: Sep 30 1997 Abstract: Some bosonic solutions of supergravities admit Killing spinors of unbroken supersymmetry. The anti-Killing spinors of broken supersymmetry can be used to generate the superpartners of stringy black holes. This has a consequent feedback on the metric and the graviphoton. The authors have found, however, that the fixed scalars for the black hole superpartners remain the same as for the original black holes. Journal volume: 56 Issue number: 6 Page number: 3509-3514 Subjects: Black Holes/;

Supergravity/spinors; Vectors; Photons; Scalars; Supergravity; Spinors; Supersymmetry; Symmetry Breaking; Sparticles; Gravitons

Kan, Nahomi; Maki, Takuya; Shiraishi, Kiyoshi Equation of state for a classical gas of BPS black holes Journal Physical Review Date: Nov 15 2001 Abstract: Abstract available at publisher web site Journal volume: 64 Issue number: 10 Page number: 104009-104009-10 Publisher: American Physical Society

Kanti, P.; Winstanley, E. Do stringy corrections stabilize colored black holes? Journal Physical Review. D, Particles Fields Date: Apr 15 2000 Abstract: The authors consider hairy black hole solutions of Einstein-Yang-Mills-dilaton theory, coupled to a Gauss-Bonnet curvature term, and the authors study their stability under small, spacetime-dependent perturbations. The authors demonstrate that stringy corrections do not remove the sphaleronic instabilities of colored black holes with the number of unstable modes being equal to the number of nodes of the background gauge function. In the gravitational sector and in the limit of an infinitely large horizon, colored black holes are also found to be unstable.

Kanti, P.; Winstanley, E. Do stringy corrections stabilize colored black holes? Journal Physical Review Date: Apr 15 2000 Abstract: Abstract available at publisher web site Journal volume: 61 Issue number: 8 Page number: 084032-084032-14 Publisher: American Physical Society

Kaplan, D.M; Strominger, A.; Lowe, D.A.; Maldacena, J.M. Microscopic entropy of N=2 extremal black holes Journal Physical Review, D (Particles Fields) Date: Apr 30 1997 Abstract: String theory is used to compute the microscopic entropy for several examples of black holes in compactifications with N=2 supersymmetry. Agreement with the Bekenstein-Hawking entropy and the moduli-independent N=2 area formula is found in all cases. Journal volume: 55 Issue number: 8 Page number: 4898-4902 Subjects: Black Holes/entropy; Quantum Gravity; String Models; Entropy; Compactification; Supersymmetry; Thermodynamics; Statistical Mechanics; Superstring Models

Kapusta, J. I. Relativistic Viscous Fluid Description of Microscopic Black Hole Wind Journal Physical Review Letters Date: Feb 26 2001 Abstract: Microscopic black holes explode with their temperature varying inversely as their mass. Such explosions would lead to the highest temperatures in the present Universe, all the way to the Planck energy. Whether or not a quasistationary shell of interacting matter undergoing radial hydrodynamic expansion surrounds such black holes is controversial. In this paper relativistic viscous fluid equations are applied to the problem assuming sufficient particle interaction. It is shown that a self-consistent picture emerges of a fluid just marginally kept in local thermal equilibrium; viscosity is a crucial element of the dynamics. Journal volume: 86 Issue number: 9 Page number: 1670-1673 Publisher: The American Physical Society

Kar, S. Stringy black holes and energy conditions Journal Physical Review, D (Particles Fields) Date: Apr 30 1997 Abstract: The energy condition inequalities for the matter stress energy comprised of the dilaton and Maxwell fields in the dilaton-Maxwell gravity theories emerging out of string theory are examined in detail. In the simplistic (1+1)-dimensional models, $R\{le\}0$ (where R is the Ricci scalar) turns out to be the requirement for ensuring focusing of timelike geodesics. In 3+1 dimensions, the authors outline the requirements on matter for pure dilaton theories these in turn constrain the functional forms of the dilaton. Journal volume: 55 Issue number: 8 Page number: 4872-4879 Subjects: Black Holes/string models; Differential Geometry; Conservation Laws; Energy; Quantum Gravity; Matter; Stresses; Energy Conservation; Space-Time

Kar, Sayan Naked singularities in low-energy, effective string theory Journal Classical and Quantum Gravity Date: Jan 01 1999 Abstract: Solutions to the equations of motion of the low-energy, effective field theory emerging out of compactified heterotic string theory are constructed by making use of the well known duality symmetries. Beginning with four-dimensional solutions of the Einstein massless scalar field theory in the canonical frame the authors first rewrite the corresponding solutions in the string frame. Thereafter, using the T- and S-duality symmetries of the low-energy string effective action the authors arrive at the corresponding uncharged, electrically charged and magnetically charged solutions. Brief comments on the construction of dual versions of the Kerr-Sen type using the dilatonic Kerr solution as the seed are also included.. Journal volume: 16 Issue number: 1 Page number: 101-115 Publisher: Institute of Physics

Kelly, Bernard; Laguna, Pablo; Lockitch, Keith; Pullin, Jorge; Schnetter, Erik; Shoemaker, Deirdre; Tiglio, Manuel Cure for unstable numerical evolutions of single black holes: Adjusting the standard ADM equations in the spherically symmetric case Journal Physical Review Date: Oct 15 2001 Abstract: Abstract available at publisher web site Journal volume: 64 Issue number: 8 Page number: 084013-084013-14 Publisher: American Physical Society

Kerimo, Johannes Dynamical black holes in scalar-tensor theories Journal Physical Review Date: Nov 15 2000 Abstract: Abstract available at publisher web site Journal volume: 62 Issue number: 10 Page number: 104005-104005-9 Publisher: American Physical Society

Keski-Vakkuri, E. Bulk and boundary dynamics in BTZ black holes Journal Physical Review, D (Particles Fields) Date: May 31 1999 Abstract: Recently, the AdS-CFT conjecture of Maldacena has been investigated in a Lorentzian signature by Balasubramanian *et al.* The authors extend this investigation to Lorentzian BTZ black hole spacetimes, and study the bulk and boundary behavior of massive scalar fields both in the nonextremal and extremal case. Using the bulk-boundary correspondence, the authors also evaluate the two-point correlator of operators coupling to the scalar field at the boundary of the spacetime, and find that it satisfies thermal periodic boundary conditions relevant to the Hawking temperature of the BTZ black hole. Journal volume: 59 Issue number: 10 Page number: 104001 Subjects: Black Holes; Space-Time; Boundary Conditions; Scalar Fields

Khanna, Gaurav Collision of spinning black holes in the close limit Journal Physical Review Date: Jun 15 2001 Abstract: In this paper the authors consider the collision of spinning holes using first order perturbation theory of black holes (Teukolsky formalism). With these results (along with ones published in the past) one can predict the properties of the gravitational waves radiated from the late stage in spiral of two spinning, equal mass black holes. Also, the authors note that the energy radiated by the head-on collision of two spinning holes with spins (that are equal and opposite) aligned along the common axis is more than the case in which the spins are perpendicular to the axis of the collision. Journal volume: 63 Issue number: 12 Page number: 124007-124007-6 Publisher: The American Physical Society

Khanna, Gaurav Collision of spinning black holes in the close limit Journal Physical Review Date: Jun 15 2001 Abstract: Abstract available at publisher web site Journal volume: 63 Issue number: 12 Page number: 124007-124007-6 Publisher: American Physical Society

Khanna, Gaurav; Baker, John; Gleiser, Reinaldo J.; Laguna, Pablo; Nicasio, Carlos O.; Nollert, Hans-Peter; Price, Richard; Pullin, Jorge Inspiraling Black Holes: The Close Limit Journal Physical Review Letters Date: Nov 01 1999 Abstract: Abstract available at publisher web site Journal volume: 83 Issue number: 18 Page number: 3581-3584 Publisher: American Physical Society

Kidder, Lawrence E.; Scheel, Mark A.; Teukolsky, Saul A.; Carlson, Eric D.; Cook, Gregory B. Black hole evolution by spectral methods Journal Physical Review Date: Oct 15 2000 Abstract: Abstract available at publisher web site Journal volume: 62 Issue number: 8 Page number: 084032-084032-20 Publisher: American Physical Society

Kim, Hee Il Primordial black holes under the double inflationary power spectrum Journal Physical Review Date: Sep 15 2000 Abstract: Abstract available at publisher web site Journal volume: 62 Issue number: 6 Page number: 063504-063504-7 Publisher: American Physical Society

Kim, Hongsu; Lee, Chul Hoon; Lee, Hyun Kyu Nonvanishing magnetic flux through the slightly charged Kerr black hole Journal Physical Review Date: Mar 15 2001 Abstract: Abstract available at publisher web site Journal volume: 63 Issue number: 6 Page number: 064037-064037-14 Publisher: American Physical Society

Kim, S.P.; Kim, S.K.; Soh, K.; Yee, J.H. Renormalized thermodynamic entropy of black holes in higher dimensions Journal Physical Review, Date: Feb 28 1997 Abstract: The authors study the ultraviolet divergent structures of the matter (scalar) field in a higher D-dimensional Reissner-Nordström black hole and compute the matter field contribution to the Bekenstein-Hawking entropy by using the Pauli-Villars regularization method. The authors find that the matter field contribution to the black hole entropy does not, in general, yield the correct renormalization of the gravitational coupling constants. Journal volume: 554 Page number: 2159-2173 Subjects: Black Holes-- Entropy; Coupling Constants; Gravitational Interactions; Many-Dimensional Calculations; Matter; Quantum Gravity; Renormalization; Scalar Fields; Thermodynamics; Basic Interactions; Field Theories; Interactions; Physical Properties; Quantum Field Theory; Thermodynamic Properties

Klebanov, I.R; Krasnitz, M. Fixed scalar greybody factors in five and four dimensions Journal Physical Review, D (Particles Fields) Date: Mar 31 1997 Abstract: The authors perform the classical gravity calculations of the fixed scalar absorption cross sections by D=5 black holes with three charges and by D=4 black holes with four charges. The authors obtain analytic results for the cases where the energy and the left- and right-moving temperatures are sufficiently low but have arbitrary ratios. Journal volume: 55 Issue number: 6 Page number: R3250-R3254 Subjects: Scalars; Absorption; Cross Sections; String Models; Coupling

Kleihaus, Burkhard; Kunz, Jutta Rotating Hairy Black Holes Journal Physical Review Letters Date: Apr 23 2001 Abstract: Abstract available at publisher web site Journal volume: 86 Issue number: 17 Page number: 3704-3707 Publisher: American Physical Society

Klemm, D.; Moretti, V.; Vanzo, L. Erratum: Rotating topological black holes [Phys. Rev. D 57, 6127 (1998)] Journal Physical Review Date: Nov 15 1999 Abstract: Abstract available at publisher web site Journal volume: 60 Issue number: 10 Page number: 109902-109902-1 Publisher: American Physical Society

Klemm, Dietmar Topological black holes in Weyl conformal gravity Journal Classical and Quantum Gravity Date: Oct 01 1998 Abstract: The authors present a class of exact solutions of Weyl conformal gravity, which have an interpretation as topological black holes. Solutions with negative, zero or positive scalar curvature at infinity are found, the former generalizing the well known topological black holes in anti-de Sitter gravity. The rather delicate question of thermodynamic properties of such objects in Weyl conformal gravity is discussed; suggesting that the thermodynamics of the solutions found should be treated within the framework of gravity as an induced phenomenon, in the spirit of Sakharov's work. Journal volume: 15 Issue number: 10 Page number: 3195-3201 Publisher: Institute of Physics

Kluson & Caron, J. Some comments about Schwarzschild black holes in matrix theory Journal Physical Review Date: May 15 2000 Abstract: Abstract available at publisher web site Journal volume: 61 Issue number: 10 Page number: 104018-104018-13 Publisher: American Physical Society

Knutt-Wehlau, M.E.; Mann, R.B. Cosmological supergravity from a massive superparticle and super-cosmological black holes Journal Classical and Quantum Gravity Date: Mar 01 1999 Abstract: The authors describe in superspace a classical theory of two-dimensional (1, 1) dilaton supergravity with a cosmological constant, both with and without coupling to a massive superparticle. The authors give general exact non-trivial superspace solutions for the compensator superfield that describes the supergravity in both cases. The authors then use these compensator solutions to construct models of two-dimensional supersymmetric cosmological black holes. Journal volume: 16 Issue number: 3 Page number: 937-952 Institute of Physics

Kohri, K.; Yokoyama, Jun'ichi Primordial black holes and primordial nucleosynthesis: Effects of hadron injection from low mass holes Journal Physical Review Date: Jan 15 2000 Abstract: Abstract available at publisher web site Journal volume: 61 Issue number: 2 Page number: 023501-023501-10 Publisher: American Physical Society

Kol, B.; Rajaraman, A. Fixed scalars and suppression of Hawking evaporation Journal Physical Review, D (Particles Fields) Date: Jul 31 1997 Abstract: For an extreme charged black hole some scalars take on a fixed value at the horizon determined by the charges alone. The authors call them fixed scalars. The authors find the absorption cross section for a low-frequency wave of a fixed scalar to be proportional to the square of the frequency. Journal volume: 56 Issue number: 2 Page number: 983-986 Subjects: Charges; Cross Sections; Radiations; Supergravity

Krasnitz, M; Klebanov, I.R. Testing effective string models of black holes with fixed scalars Journal Physical Review, D (Particles Fields) Date: Aug 31 1997 Abstract: The authors solve the problem of mixing between the fixed scalar and metric fluctuations. First, the authors derive the decoupled fixed scalar equation for the four-dimensional black hole with two different charges. The authors proceed to the five-dimensional black hole with different electric (one-brane) and magnetic (five-brane) charges, and derive two decoupled equations satisfied by appropriate mixtures of the original fixed scalar fields. Journal volume: 56 Issue number: 4 Page number: 2173-2179 Subjects: Scalar Fields; Mixing; Fluctuations; Coupling; Chiral Symmetry; Absorption; Cross Sections

Krasnov, K. Quanta of geometry and rotating black holes Journal Classical and Quantum Gravity Date: Apr 01 1999 Abstract: In the loop approach to quantum gravity the spectra of operators corresponding to such geometrical quantities as length, area and volume become quantized. However, the size of arising quanta of geometry in Planck units is not fixed by the theory itself: a free parameter, sometimes referred to as the Immirzi parameter, is known to affect the spectrum of all geometrical operators. In this paper I propose an argument that fixes the value of this parameter. I consider rotating black holes, in particular the extremal ones. For such black holes the 'no naked singularity condition' bounds the total angular momentum J by , where is the horizon area and G is Newton's constant. A similar bound on J comes from the quantum theory. The requirement that these two bounds are the same fixes the value of the Immirzi parameter to be unity. A byproduct of this argument is a picture of the quantum extremal rotating black hole in which all the spin entering the extremal hole is concentrated in a single puncture. Journal volume: 16 Issue number: 4 Page number: L15-L18 Publisher: Institute of Physics

Krasnov, K. The area spectrum in quantum gravity Journal Classical and Quantum Gravity Date: Jun 01 1998 Abstract: The authors show that, apart from the usual area operator of non-perturbative quantum gravity, there exists another, closely related, operator that measures areas of surfaces. Both corresponding classical expressions yield the area. Quantum mechanically, however, the spectra of the two operators are different, coinciding only in the limit when the spins labelling the state are large. The authors argue that both operators are legitimate quantum operators, and the choice of which one to use depends on the context of the physical problem of interest.

Krasnov, Kirill V. Quantum geometry and thermal radiation from black holes Journal Classical and Quantum Gravity Date: Feb 01 1999 Abstract: A quantum mechanical description of black hole states proposed recently within non-perturbative quantum gravity is used to study the emission and absorption spectra of quantum black holes. The authors assume that the probability distribution of states of the quantum black hole is given by the 'areá canonical ensemble, in which the horizon area is used instead of energy, and the authors use Fermi's golden rule to find the line intensities. For a non-rotating black hole, the authors study the absorption and emission of s-waves considering a special set of emission lines. To find the line intensities the authors use an analogy between a microscopic state of the black hole and a state of the gas of atoms. Journal volume: 16 Issue number: 2 Page number: 563-578 Publisher: Institute of Physics

Krennrich, F.; Le Bohec, S.; Weekes, T. C. Detection Techniques of Microsecond Gamma-Ray Bursts Using Ground-based Telescopes Journal Astrophysical Journal Date: Jan 20 2000 Abstract: Gamma-ray observations above 200 MeV are conventionally made by satellite-based detectors. The EGRET detector on the Compton Gamma Ray Observatory has provided good sensitivity for the detection of bursts lasting for more than 200 ms. Theoretical predictions of high-energy gamma-ray bursts produced by quantum mechanical decay of primordial black holes (Hawking) suggest the emission of bursts on shorter time scales. The final stage of a primordial black hole results in a burst of gamma rays, peaking around 250 MeV and lasting for 1/10 of a microsecond or longer depending on particle physics. In this work the authors show that there is an observational window using ground-based imaging Cerenkov detectors to measure gamma-ray burst emission at energies $E>200$ MeV. This technique, with a sensitivity for bursts lasting nanoseconds to several microseconds, is based on the detection of multiphoton-initiated air showers. (c) (c) 2000. The American Astronomical Society. Journal volume: 529 Issue number: 1 Page number: 506-512

Kribs, G.D.; Leibovich, A.K.; Rothstein, I.Z. Bounds from primordial black holes with a near critical collapse initial mass function Journal Physical Review, Date: Nov 01 1999 Abstract: Recent numerical evidence suggests that a mass spectrum of primordial black holes (PBHs) is produced as a consequence of near critical gravitational collapse. Assuming that these holes formed from the initial density perturbations seeded by inflation, the authors calculate model independent upper bounds on the mass variance at the reheating temperature by requiring that the mass density does not exceed the critical density and the photon emission does not exceed current diffuse gamma-ray measurements.

Kribs, Graham D.; Leibovich, Adam K.; Rothstein, I. Z. Bounds from primordial black holes with a near critical collapse initial mass function Journal Physical Review Date: Nov 15 1999 Abstract: Abstract available at publisher web site

Kukula, Marek J.; Dunlop, James S.; McLure, Ross J.; Miller, Lance; Percival, Will J.; Baum, Stefi A.; O'Dea, Christopher P. A NICMOS imaging study of high-z quasar host galaxies Journal Monthly Notices of the Royal Astronomical Society Date: Oct 01 2001 Abstract: The authors present the first results from a major Hubble Space Telescope programme designed to investigate the cosmological evolution of quasar host galaxies from &formmu1; to the present day.

Kummer, W.; Vassilevich, D. V. Effective action and Hawking radiation for dilaton coupled scalars in two dimensions Journal Physical Review Date: Oct 15 1999 Abstract: Abstract available at publisher web site Journal volume: 60 Issue number: 8 Page number: 084021-084021-8 Publisher: American Physical Society

Kummer, W; Lau, S.R. Boundary conditions and quasilocal energy in the canonical formulation of all 1+1 models of gravity Journal Annals of Physics (New York) Date: Jul 31 1997 Abstract: Within a first-order framework, the authors comprehensively examine the role played by boundary conditions in the canonical formulation of a completely general two-dimensional gravity model. Our analysis particularly elucidates the perennial themes of mass and energy. The gravity models for which our arguments are valid include theories with dynamical torsion and so-called generalized dilaton theories (GDTs). Journal volume: 258 Issue number: 1 Page number: 37-80 Subjects: General Relativity Theory; Mass; Action Integral; Space-Time; Energy-Momentum Tensor

Kupi, Gabor Interaction between black holes and dark matter Journal Physical Review Date: Nov 15 2001 Abstract: Abstract available at publisher web site Journal volume: 64 Issue number: 10 Page number: 103507-103507-4 Publisher: American Physical Society

Larsen, A.L.; Lousto, C.O. Are higher order membranes stable in black hole spacetimes? Journal Physical Review, D (Particles Fields) Date: Jun 30 1997 Abstract: The authors continue the study of the existence and stability of static spherical membrane configurations in curved spacetimes. The authors first consider higher order membranes described by a Lagrangian which, in addition to the Dirac term, includes a term proportional to the scalar curvature of the world volume $\{sup(3)\}R$. Journal volume: 55 Issue number: 12 Page number: 7936-7941 Subjects: Lagrangian Function; Equations Of Motion

Larsen, F. String model of black hole microstates Journal Physical Review, D (Particles Fields) Date: Jul 31 1997 Abstract: The statistical mechanics of black holes arbitrarily far from extremality is modeled by a gas of weakly interacting strings. As an effective low-energy description of black holes the string model provides several highly nontrivial consistency checks and predictions. Journal volume: 56 Issue number: 2 Page number: 1005-1008 Subjects: Statistical Mechanics; Gases; Thermodynamics

Larsen, F.; Wilczek, F. Resolution of cosmological singularities in string theory Journal Physical Review, D (Particles Fields) Date: Apr 30 1997 Abstract: The authors show that a class of (3+1)-dimensional Friedmann-Robertson-Walker cosmologies can be embedded within a variety of solutions of string theory. In some realizations, the apparent singularities associated with the big bang or big crunch are resolved at nonsingular horizons of higher-dimensional quasi-black hole solutions (with compactified *real* time); in others, plausibly they are resolved at D-brane bound states having no conventional space-time interpretation. Journal volume: 55 Issue number: 8 Page number: 4591-4595 Subjects: Cosmology; Singularity; Four-Dimensional

Calculations; String Models; Bound State

Larsen, F.; Wilczek, F. Resolution of cosmological singularities in string theory Journal Physical Review, Date: Apr 30 1997 Abstract: The authors show that a class of (3+1)-dimensional Friedmann-Robertson-Walker cosmologies can be embedded within a variety of solutions of string theory. In some realizations, the apparent singularities associated with the big bang or big crunch are resolved at nonsingular horizons of higher-dimensional quasi-black hole solutions (with compactified *real time*); in others, plausibly they are resolved at D-brane bound states having no conventional space-time interpretation. Journal volume: 558 Page number: 4591-4595 Subject keyword: General Relativity Theory; Extended Particle Model; Field Theories; Mathematical Models; Particle Models

Lasota, Jean-Pierre; Abramowicz, Marek A. Cet objet obscur: le trou noir Journal Classical and Quantum Gravity Date: Jan 01 1997 Abstract: Advection-dominated accretion flows could be a unique signature of the presence of black holes in various accreting astrophysical systems such as some quiescent transient x-ray sources and low-luminosity nuclei of galaxies. The authors present the general framework describing such advection-dominated flows around Kerr black holes and point out several problems that remain to be solved. Journal volume: 14 Issue number: 1A Page number: A237-A250 Publisher: Institute of Physics

Leach, Samuel M.; Grivell, Ian J.; Liddle, Andrew R. Black hole constraints on the running-mass inflation model Journal Physical Review Date: Aug 15 2000 Abstract: Abstract available at publisher web site Journal volume: 62 Issue number: 4 Page number: 043516-043516-6 Publisher: American Physical Society

Lee, H.W.; Kim, N.J.; Myung, Y.S. Dilaton test of connection between AdS $3 \times S 3$ and 5D black holes Journal Classical and Quantum Gravity Date: Jan 07 2000 Abstract: A five-dimensional black hole (M 5) is investigated in the type IIB superstring theory compactified on $S 1 \times T 4$. This corresponds to AdS $3 \times S 3 \times T 4$ in the near horizon with asymptotically flat space. Here the harmonic gauge is introduced to decouple the mixing between the dilaton and others. On the other hand, the authors obtain the BTZ black hole (AdS $3 \times S 3 \times T 4$) as the non-dilatonic solution. The authors calculate the grey-body factor of the dilaton as a test scalar both for a 5D black hole (M $5 \times S 1 \times T 4$) and the BTZ black hole (AdS $3 \times S 3 \times T 4$). The result of the BTZ black hole agrees with the grey-body factor of the dilaton in the dilute gas approximation of a 5D black hole. Journal volume: 17 Issue number: 1 Page number: 113-122 Publisher: Institute of Physics

Lee, H.W.; Myung, Y.S.; Kim, Jin Young The classical stability of charged (1 + 1)-dimensional black holes Journal Classical and Quantum Gravity Date: Mar 01 1997 Abstract: The authors discuss the classical stability of charged, two-dimensional, black holes. Here is the dilaton - Maxwell coupling parameter. The authors introduce a tachyon field, which arises from string theories, as a test field. One finds barrier - well type potentials surrounding the extremal black holes, which induce the classical instability. For the non-extremal black holes, one finds the barrier - well type potentials between the inner (Cauchy) horizon and outer (event) horizon, while potential barriers appear outside the event horizon. This implies that the inner horizons are unstable and the outer horizons are stable. However, this means that the non-extremal black holes themselves are stable because the stability of a black hole depends only on the outer horizon. Finally the authors mention that the instability of extremal black holes may originate from the unstable inner horizon of non-extremal black holes. Journal volume: 14 Issue number: 3 Page number: 759-774 Publisher: Institute of Physics

Lee, Hyun Kyu; Lee, Chul H.; van Putten, Maurice H. P. M. Electric charge and magnetic flux on rotating black holes in a force-free magnetosphere Journal Monthly Notices of the Royal Astronomical Society Date: Jun 01 2001 Abstract: The electric charge on rotating black holes is calculated to be ~BJ in the force-free configuration of Ghosh, with a horizon flux of ~BM 2. This charge is gravitationally weak for B~1015 G, so that the Kerr metric applies. Being similar to the electric charge of a magnetar, both electric charge and magnetic flux should be, in sign and order of magnitude, continuous during stellar collapse into a black hole. Extraction of the rotational energy from newly formed black holes may proceed by interaction with the magnetic field. Journal volume: 324 Issue number: 3 Page number: 781-784 DOE sponsorship Publisher: Blackwell Science

Lee, Hyung Mok Growth of stellar mass black holes in galactic nuclei Journal Classical and Quantum Gravity Date: Oct 07 2001 Abstract: P The authors have examined the dynamical evolution of dense stellar systems containing black holes that are formed by evolution of massive stars. The black holes, whose mass is approximately 10 M , become the most massive component in a few times 107 yr. They quickly form a subsystem of very small size within the core of the cluster of ordinary stars via dynamical friction. The subsystem evolves on its own relaxation time scale, which is much shorter than the dynamical evolution time scale of the surrounding system.

Lee, William H. Newtonian hydrodynamics of the coalescence of black holes with neutron stars - III. Irrotational binaries with a stiff equation of state Journal Monthly Notices of the Royal Astronomical Society Date: Oct 01 2000 Abstract: The authors present a numerical study of the hydrodynamics in the final stages of inspiral in a black hole-neutron star binary, when the binary separation becomes comparable to the stellar radius. The authors use a Newtonian three-dimensional Smooth Particle Hydrodynamics (SPH) code, and model the neutron star with a stiff (adiabatic index $\Gamma=3$ and 2.5) polytropic equation of state and the black hole as a Newtonian point mass which accretes matter via an absorbing boundary at the Schwarzschild radius

Lee, William H. Newtonian hydrodynamics of the coalescence of black holes with neutron stars - IV. Irrotational binaries with a soft equation of state Journal Monthly Notices of the Royal Astronomical Society Date: Dec 01 2001 Abstract: The authors present the results of three-dimensional hydrodynamical simulations of the final stages of in-spiral in a black hole-neutron star binary, when the separation is comparable to the stellar radius. The authors use a Newtonian smooth particle hydrodynamics (SPH) code to model the evolution of the system.

Letelier, P.S.; Vieira, W.M. Chaos in black holes surrounded by gravitational waves Journal Classical and Quantum Gravity Date: May 01 1997 Abstract: The occurrence of chaos for test particles moving around a Schwarzschild black hole perturbed by a special class of gravitational waves is studied in the context of the Melnikov method. The explicit integration of the equations of motion for the homoclinic orbit is used to reduce the application of this method to the study of simple graphs. Journal volume: 14 Issue number: 5 Page number: 1249-1257 Publisher: Institute of Physics

Letelier, Patricio S.; Oliveira, Samuel R. Uniformly accelerated black holes Journal Physical Review Date: Sep 15 2001 Abstract: Abstract available at publisher web site radicand: 27) is just a coordinate artifact. The main results are extended to accelerated rotating black holes with no significant changes. Journal volume: 64 Issue number: 6 Page number: 064005-064005-9 Publisher: American Physical Society

Levinson, Amir Particle Acceleration and Curvature TeV Emission by Rotating, Supermassive Black Holes Journal Physical Review Letters Date: Jul 31 2000 Abstract: Abstract available at publisher web site Journal volume: 85 Issue number: 5 Page number: 912-915 Publisher: American Physical Society

Levinson, Amir; Waxman, Eli Probing Microquasars with TeV Neutrinos Journal Physical Review Letters Date: Oct 22 2001 Abstract: Abstract available at publisher web site Journal volume: 87 Issue number: 17 Page number: 171101-171101-4 Publisher: American Physical Society

Li, M; Martinec, E; Sahakian, V. Black holes and the SYM phase diagram Journal Physical Review, D (Particles Fields) Date: Feb 28 1999 Abstract: Making combined use of the matrix and Maldacena conjectures, the relation between various thermodynamic transitions in super Yang-Mills (SYM) theory and supergravity is clarified. The thermodynamic phase diagram of an object in DLCQ M theory in four and five non-compact space dimensions is constructed; matrix strings, matrix black holes, and black p-branes are among the various phases. Critical manifolds are characterized by the principles of correspondence and longitudinal localization, and a triple point is identified. Journal volume: 59 Issue number: 4 Page number: 044035 Subjects: Supersymmetry; Black Holes; Supergravity; Yang-Mills Theory; Thermodynamics; Matrices; String Models; Phase Transformations

Li, Miao; Martinec, Emil Matrix black holes Journal Classical and Quantum Gravity Date: Dec 01 1997 Abstract: Four- and five-dimensional extremal black holes with nonzero entropy have simple presentations in M-theory as gravitational waves bound to configurations of intersecting M-branes. The authors discuss realizations of these objects in matrix models of M-theory, investigate the properties of 0-brane probes, and propose a measure of their internal density. A scenario for black-hole dynamics is presented. Journal volume: 14 Issue number: 12 Page number: 3187-3204 Publisher: Institute of Physics

Li, Miao; Martinec, Emil On the entropy of matrix black holes Journal Classical and Quantum Gravity Date: Dec 01 1997 Abstract: The authors compute the entropy of 5D black holes carrying up to three charges using matrix theory. Journal volume: 14 Issue number: 12 Page number: 3205-3213 Publisher: Institute of Physics

Li, Zhong-heng Quantum corrections to the entropy of a Reissner-Nordström black hole due to spin fields Journal Physical Review Date: Jul 15 2000 Abstract: Abstract available at publisher web site Journal volume: 62 Issue number: 2 Page number: 024001-024001-3 Publisher: American Physical Society

Liang, Y. C.; Teo, Edward Black diholes with unbalanced magnetic charges Journal Physical Review Date: Jul 15 2001 Abstract: Abstract available at publisher web site Journal volume: 64 Issue number: 2 Page number: 024019-024019-7 Publisher: American Physical Society

Liberati, Stefano; Rothman, Tony; Sonego, Sebastiano Nonthermal nature of incipient extremal black holes Journal Physical Review Date: Jul 15 2000 Abstract: Abstract available at publisher web site Journal volume: 62 Issue number: 2 Page number: 024005-024005-10 Publisher: American Physical Society

Liberati, Stefano; Rothman, Tony; Sonego, Sebastiano; Ahluwalia, D. V. Extremal Black Holes and the Limits of the Third Law Journal International Journal of Modern Physics Date: Feb 01 2001 Abstract: Recent results of quantum field theory on a curved spacetime suggest that extremal black holes are not thermal objects and that the notion of zero temperature is ill-defined for them. If this is correct, one may have to go to a full semiclassical theory of gravity, including back reaction, in order to make sense of the third law of black hole thermodynamics. Alternatively it is possible that the authors shall have to drastically revise the status of extremality in black hole thermodynamics. Journal volume: 10 Issue number: 01

Rights: World Scientific Publishing Company Publisher: World Scientific Publishing Company

Lin, Feng-Li; Soo, Chopin Quantum field theory with and without conical singularities: black holes with a cosmological constant and the multi-horizon scenario Journal Classical and Quantum Gravity Date: Feb 01 1999 Abstract: Boundary conditions and the corresponding states of a quantum field theory depend on how the horizons are taken into account. There is an ambiguity as to which method is appropriate because different ways of incorporating the horizons lead to different results. The authors propose that a natural way of including the horizons is to first consider the Kruskal extension and then define the quantum field theory on the Euclidean section.

Linet, B. Entropy bound of a charged object and electrostatic self-energy in black holes Journal Physical Review Date: May 15 2000 Abstract: Abstract available at publisher web site Journal volume: 61 Issue number: 10 Page number: 107502-107502-4 Publisher: American Physical Society

Liu, Hongya; Wesson, Paul S. The physical properties of charged five-dimensional black holes Journal Classical and Quantum Gravity Date: Jul 01 1997 Abstract: The authors give a class of exact solutions of the five-dimensional (Kaluza - Klein) equations of general relativity, identify the mass and electric charge of the source and solve the geodesic equation for the motion of a charged test particle. There are interesting differences between charged black holes in four and five dimensions that could, in principle, be used to test the dimensionality of the world. Journal volume: 14 Issue number: 7 Page number: 1651-1663 Publisher: Institute of Physics

Liu, Y; Hu, L; Ge, M. Effect of violation of quantum mechanics on neutrino oscillation Journal Physical Review, D (Particles Fields) Date: Nov 30 1997 Abstract: The effect of quantum mechanics violation due to quantum gravity on neutrino oscillation is investigated. It is found that the mechanism introduced by Ellis, Hagelin, Nanopoulos, and Srednicki through the modification of the Liouville equation can affect neutrino oscillation behavior and may be taken as a new solution of the solar neutrino problem. Journal volume: 56 Issue number: 10 Page number: 6648-6652 Subjects: Neutrino Oscillation/Quantum Mechanics; Violations; Quantum Gravity; Solar Neutrinos;Fluctuations

Lousto, Carlos O. Perturbative evolution of nonlinear initial data for binary black holes: Zerilli versus Teukolsky equation Journal Physical Review Date: Feb 15 2001 Abstract: Abstract available at publisher web site Journal volume: 63 Issue number: 4 Page number: 047504-047504-4 Publisher: American Physical Society

Lousto, Carlos O. Pragmatic Approach to Gravitational Radiation Reaction in Binary Black Holes Journal Physical Review Letters Date: Jun 05 2000 Abstract: Abstract available at publisher web site Journal volume: 84 Issue number: 23 Page number: 5251-5254 Publisher: American Physical Society

Lousto, Carlos O. Towards the solution of the relativistic gravitational radiation reaction problem for binary black holes Journal Classical and Quantum Gravity Date: Oct 07 2001 Abstract: P Here the authors present the results of applying the generalized Riemann ζ-function regularization method to the gravitational radiation reaction problem. The authors analyse in detail the head-on collision of two non-spinning black holes with an extreme mass ratio. The resulting reaction force on the smaller hole is repulsive. The authors discuss the possible extensions of these method to generic orbits and spinning black holes. The determination of corrected trajectories allows us to add second perturbative corrections with the consequent increase in the accuracy of computed waveforms. Journal volume: 18

Issue number: 19 Page number: 3989-3994 Publisher: Institute of Physics

Lowe, D.A. Statistical Origin of Black Hole Entropy in Matrix Theory Journal Physical Review Letters Date: Jul 31 1998 Abstract: The statistical entropy of black holes in matrix theory is considered. Assuming matrix theory is the discretized light-cone quantization of a theory with eleven-dimensional Lorentz invariance, the authors map the counting problem onto the original Gibbons-Hawking calculations of the thermodynamic entropy. Journal volume: 81 Issue number: 2 Page number: 256-259 Subjects: Matrix Elements; Light Cone; Lorentz Invariance; Entropy; Thermodynamics; Supersymmetry

Lowe, D.A.; Thorlacius, L. AdS-CFT correspondence and the information paradox Journal Physical Review, Date: Nov 01 1999 Abstract: The information paradox in the quantum evolution of black holes is studied within the framework of the anti–de Sitter–conformal field theory (AdS-CFT) correspondence. The unitarity of the CFT strongly suggests that all information about an initial state that forms a black hole is returned in the Hawking radiation. The CFT dynamics implies an information retention time of the order of the black-hole lifetime. This fact determines many qualitative properties of the nonlocal effects that must show up in a semiclassical effective theory in the bulk.

Lowe, David A. Comment on "Do Semiclassical Zero Temperature Black Holes Exist" Journal Physical Review Letters Date: Jul 09 2001 Abstract: Abstract available at publisher web site Journal volume: 87 Issue number: 2 Page number: 029001-029001-1 Publisher: American Physical Society

Lu, Youjun; Yu, Qingjuan The relationship between X-ray variability and the central black hole mass Journal Monthly Notices of the Royal Astronomical Society Date: Jun 01 2001 Abstract: The authors assembled a sample of Seyfert 1 galaxies, quasi-stellar objects (QSOs) and low-luminosity active galactic nuclei (LLAGNs) observed by ASCA, the central black hole masses of which have been measured. The authors found that the X-ray variability (which is quantified by the 'excess variance'

Lubo, M. Quantum minimal length and trans-Planckian photons Journal Physical Review Date: Jun 15 2000 Abstract: Abstract available at publisher web site Journal volume: 61 Issue number: 12 Page number: 124009-124009-11 Publisher: American Physical Society

Lue, Arthur; Weinberg, Erick J. Gravitational properties of monopole spacetimes near the black hole threshold Journal Physical Review. D, Particles Fields Date: Jun 15 2000 Abstract: Although nonsingular spacetimes and those containing black holes are qualitatively quite different, there are continuous families of configurations that connect the two. In this paper the authors use self-gravitating monopole solutions as tools for investigating the transition between these two types of spacetimes. The authors show how causally distinct regions emerge as the black hole limit is achieved, even though the measurements made by an external observer vary continuously. The authors find that near-critical solutions have a naturally defined entropy, despite the absence of a true horizon, and that this has a clear connection with the Hawking-Bekenstein entropy. Journal volume: 61 Issue number: 12 Page number: 124003-124003-10 Publisher: American Physical Society

MacGibbon, J.H.; Brandenberger, R.H; Wichoski, U.F. Limits on black hole formation from cosmic string loops Journal Physical Review, D (Particles Fields) Date: Feb 28 1998 Abstract: In theories with cosmic strings, a small fraction of string loops may collapse to form black holes. In this paper, various constraints on such models involving black holes are considered. Hawking radiation from black holes, gamma and cosmic ray flux limits and constraints from the possible formation of stable black hole remnants are reanalyzed. Journal volume: 57 Issue number: 4 Page number: 2158-2165 Subjects: Quantum Gravity; Black Holes; String Models; Cosmology; Cosmic Ray Flux; Anisotropy; Microwave Radiation; Background Radiation

Maciejewski, Witold; Binney, James Kinematics from spectroscopy with a wide slit: detecting black holes in galaxy centres Journal Monthly Notices of the Royal Astronomical Society Date: May 01 2001 Abstract: The authors consider long-slit emission-line spectra of galactic nuclei when the slit is wider than the instrumental point spread function, and the target has large velocity gradients. The finite width of the slit generates complex distributions of brightness at a given spatial point in the measured spectrum, which can be misinterpreted as coming from additional physically distinct nuclear components. The authors illustrate this phenomenon for the case of a thin disc in circular motion around a nuclear black hole (BH). The authors develop a new method for estimating the mass of the BH that exploits a feature in the spectrum at the outer edge of the BH's sphere of influence, and therefore gives higher sensitivity to BH detection than traditional methods.

Maeda, Kengo; Torii, Takashi; Narita, Makoto String excitation inside generic black holes Journal Physical Review Date: Jan 15 2000 Abstract: Abstract available at publisher web site Journal volume: 61 Issue number: 2 Page number: 024020-024020-8 Publisher: American Physical Society

Magorrian, John; Tremaine, Scott Rates of tidal disruption of stars by massive central black holes Journal Monthly Notices of the Royal Astronomical Society Date: Oct 01 1999 Abstract: There is strong evidence for some kind of massive dark object in the centres of many galaxy bulges. The detection of flares from tidally disrupted stars could confirm that these objects are black holes (BHs). Here the authors present calculations of the stellar disruption rates in detailed dynamical models of real galaxies, taking into account the refilling of the loss cone of stars on disruptable orbits by two-body relaxation and tidal forces in non-spherical galaxies.

Magueijo, João Stars and black holes in varying speed of light theories Journal Physical Review Date: Feb 15 2001 Abstract: Abstract available at publisher web site Journal volume: 63 Issue number: 4 Page number: 043502-043502-14 Publisher: American Physical Society

Mahapatra, S. String propagation in an exact four-dimensional black hole background Journal Physical Review, D (Particles Fields) Date: May 31 1997 Abstract: The authors study string propagation in an exact, stringy, four-dimensional dyonic black hole background. The exact solutions in terms of elliptic functions describing string configurations in the $J=0$ limit are obtained by solving the string equations of motion and constraints. Journal volume: 55 Issue number: 10 Page number: 6403-6408 Subjects: String Models/equations of motion; Black Holes; Dyons; Space-Time; Partial Differential Equations

Mäkelä, Jarmo; Repo, Pasi Microscopic black-hole pairs in highly excited states Journal Classical and Quantum Gravity Date: Feb 07 2001 Abstract: P The authors consider the quantum mechanics of a system consisting of two identical, Planck-sized Schwarzschild black holes revolving around their common centre of mass. The authors find that even in a very highly excited state such a system has very sharp, discrete energy eigenstates, and the system performs very rapid transitions from one stationary state to another. For instance, when the system is in the 100th excited state, the lifetimes of the energy eigenstates are of the order of 10-30 s, and the energies of gravitons released in transitions between nearby states are of the order of 1022 eV. Journal volume: 18 Issue number: 3 Page number: 373-394 Publisher: Institute of Physics

Mäkelä, Jarmo; Repo, Pasi; Luomajoki, Markus; Piilonen, Johanna Quantum-mechanical model of the Kerr-Newman black hole Journal Physical Review Date: Jul 15 2001 Abstract: Abstract available at publisher web site radicand: 2n, where n is an integer. It turns out that this result is closely related to Bekenstein's proposal on the discrete horizon area spectrum of black holes. Journal volume: 64 Issue number: 2 Page number: 024018-024018-22 Publisher: American Physical Society

Makishima, Kazuo Magnetohydrodynamic interpretations of high-energy phenomena in galaxies and clusters Journal Plasma Physics and Controlled Fusion Date: May 01 1997 Abstract: Energetic plasmas are found in all astrophysical environments. However, beyond the solar system, our understanding of cosmic plasma dynamics has remained quite poor. This paper presents an attempt to explore a new description of various high-energy astrophysical phenomena, observed from galaxies and galaxy clusters, in terms of plasma magnetohydrodynamics. In this context, three specific topics are described; diffuse hot plasmas filling the Milky Way, plasma acceleration by giant black holes, and hot plasmas of galaxy clusters with relevance to cosmology. Journal volume: 39 Issue number: 5A Page number: A15-A21 Publisher: Institute of Physics

Maldacena, J. D-branes and near extremal black holes at low energies Journal Physical Review, D (Particles Fields) Date: Jun 30 1997 Abstract: It has been observed recently that many properties of some near extremal black holes can be described in terms of bound states of D-branes. Using a nonrenormalization theorem the authors argue that the D-brane description is the correct quantum gravity description of the black hole at low energies. The low-energy theory includes the black hole degrees of freedom that account for the entropy and describes also Hawking radiation. Journal volume: 55 Issue number: 12 Page number: 7645-7650 Subjects: Entropy; Solitons; Renormalization; Quantum Gravity; Radiations; String Models

Maldacena, J. Probing near extremal black holes with D-branes Journal Physical Review, (Particles Fields) Date: Mar 31 1998 Abstract: The authors calculate the one loop effective action for D-brane probes moving in the presence of near Bogomol'nyi-Prasad-Sommerfield D-branes. The $v\{sup 2\}$ term agrees with supergravity in all cases and the static force agrees for a five-dimensional black hole with two large charges. It also agrees qualitatively in all the other cases. The authors make some comments on the M(atrix) theory interpretation of these results. Journal volume: 57 Issue number: 6 Page number: 3736-3741 Subjects: Black Holes; Probes; Solitons; Supergravity; Many-Dimensional Calculations; Matrix Elements

Maldacena, J.; Strominger, A. Universal low-energy dynamics for rotating black holes Journal Physical Review, D (Particles Fields) Date: Oct 31 1997 Abstract: Fundamental string theory has been used to show that low-energy excitations of certain black holes are described by a two-dimensional conformal field theory. This picture has been found to be extremely robust. In this paper it is argued that many essential features of the low-energy effective theory can be inferred directly from a semiclassical analysis of the general Kerr-Newman solution of supersymmetric four-dimensional Einstein-Maxwell gravity, without using string theory. Journal volume: 56 Issue number: 8 Page number: 4975-4983 Subjects: Black Holes/string models; Quantum Field Theory/Conformal Invariance; Quantum Gravity; Rotation; Semiclassical Approximation; Supersymmetry; Einstein-Maxwell Equations; Angular Momentum; Scalars; Emission; Absorption; Excitation

Malzac, Julien; Beloborodov, Andrei M.; Poutanen, Juri X-ray spectra of accretion discs with dynamic coronae Journal Monthly Notices of the Royal Astronomical Society Date: Sep 01 2001 Abstract: The authors compute the X-ray spectra produced by non-static coronae atop accretion discs around black holes and neutron stars. The hot corona is radiatively coupled to the underlying disc (the reflector) and generates an X-ray spectrum which is sensitive to the bulk velocity of the coronal plasma. The authors show that an outflowing corona reproduces the hard-state spectrum of Cyg X-1 and similar objects.

Manko, Vladimir S.; Ruiz, Eduardo An exact solution of the double-Kerr equilibrium problem Journal Classical and Quantum Gravity Date: Jan 21 2001 Abstract: P The authors present concise analytic expressions for the Komar masses and angular momenta of the balancing Kerr-NUT constituents in the double-Kerr solution which complement the equilibrium formulae previously obtained (Manko V S, Ruiz E and Sanabria-Gómez J D 2000 Class. Quantum Grav.17 3881). A rigorous proof of the non-existence of equilibrium states of two Kerr black holes with positive masses is given. Journal volume: 18 Issue number: 2 Page number: L11-L15 Publisher: Institute of Physics

Manko, Vladimir S.; Ruiz, Eduardo; Manko, Olga V. Is Equilibrium of Aligned Kerr Black Holes Possible? Journal Physical Review Letters Date: Dec 25 2000 Abstract: Abstract available at publisher web site Journal volume: 85 Issue number: 26 Page number: 5504-5506 Publisher: American Physical Society

Mann, R.B. Pair production of topological anti-de Sitter black holes Journal Classical and Quantum Gravity Date: May 01 1997 Abstract: The pair creation of black holes with event horizons of non-trivial topology is described. The space times are all limiting cases of the cosmological C-metric. They are generalizations of the (2 + 1)-dimensional black hole and have asymptotically anti-de Sitter behaviour. Domain wall instantons can mediate their pair creation for a wide range of mass and charge. Journal volume: 14 Issue number: 5 Page number: L109-L114 Publisher: Institute of Physics

Mann, Robert Black holes of negative mass Journal Classical and Quantum Gravity Date: Oct 01 1997 Abstract: I demonstrate that, under certain circumstances, regions of negative energy density can undergo gravitational collapse into a black hole. The resultant exterior black hole spacetimes necessarily have negative mass and non-trivial topology. A full theory of quantum gravity, in which topology-changing processes take place, could give rise to such spacetimes. Journal volume: 14 Issue number: 10 Page number: 2927-2930 Publisher: Institute of Physics

Mao, Shude; Witt, Hans J.; Koopmans, Leon V. E. The influence of central black holes on gravitational lenses Journal Monthly Notices of the Royal Astronomical Society Date: May 01 2001 Abstract: Recent observations indicate that many if not all galaxies host massive central black holes. In this paper the authors explore the influence of black holes on the lensing properties. The authors model the lens as an isothermal ellipsoid with a finite core radius plus a central black hole. The authors show that the presence of the black hole substantially changes the critical curves and caustics. If the black hole mass is above a critical value, then it will completely suppress the central images for all source positions. Realistic central black holes are likely to have masses below this critical value. Even in such subcritical cases, the black hole can suppress the central image when the source is inside a zone of influence, which depends on the core radius and black hole mass. In the subcritical cases, an additional image may be created by the black hole in some regions, which for some radio lenses may be detectable with high-resolution and large dynamic range VLBI maps. The presence of central black holes should also be taken into account when one constrains the core radius from the lack of central images in gravitational lenses. Journal volume: 323 Issue number: 2 Page number: 301-307 DOE sponsorship Publisher: Blackwell Science

Marolf, Donald; Trodden, Mark Black holes and instabilities of negative tension branes Journal Physical Review Date: Sep 15 2001 Abstract: Abstract available at publisher web site Journal volume: 64 Issue number: 6 Page number: 065019-065019-8 Publisher: American Physical Society

Marronetti, Pedro; Huq, Mijan; Laguna, Pablo; Lehner, Luis; Matzner, Richard A.; Shoemaker, Deirdre Approximate analytical solutions to the initial data problem of black hole binary systems Journal Physical Review Date: Jul 15 2000 Abstract: Abstract available at publisher web site Journal volume: 62 Issue number: 2 Page number: 024017-024017-5 Publisher: American Physical Society

Marronetti, Pedro; Matzner, Richard A. Solving the Initial Value Problem of Two Black Holes Journal Physical Review Letters Date: Dec 25 2000 Abstract: Abstract available at publisher web site Journal volume: 85 Issue number: 26 Page number: 5500-5503 Publisher: American Physical Society

Marsh, T.R. Gravitational lensing in eclipsing binary stars Journal Monthly Notices of the Royal Astronomical Society Date: Jun 01 2001 Abstract: I consider the effect of the gravitational deflection of light upon the light curves of eclipsing binary stars, focusing mainly upon systems containing at least one white dwarf component. In absolute terms the effects are small, however they are strongest at the time of secondary eclipse when the white dwarf transits its companion, and act to reduce the depth of this feature.

Martínez, Cristián; Teitelboim, Claudio; Zanelli, Jorge Charged rotating black hole in three spacetime dimensions Journal Physical Review Date: May 15 2000 Abstract: Abstract available at publisher web site Journal volume: 61 Issue number: 10 Page number: 104013-104013-8 Publisher: American Physical Society

Martinec, E; Sahakian, V. Black holes and the SYM phase diagram. II Journal Physical Review, D (Particles Fields) Date: Jun 30 1999 Abstract: The complete phase diagram of objects in M theory compactified on tori $T\{sup\ p\}, p=1,2,3$, is elaborated. Phase transitions occur when the object localizes on cycle(s) (the Gregory-Laflamme transition), or when the area of the localized part of the horizon becomes one in string units (the Horowitz-Polchinski correspondence point). The low-energy, near-horizon geometry that governs a given phase can match onto a variety of asymptotic regimes. The analysis makes it clear that the matrix conjecture is a special case of the Maldacena conjecture. Journal volume: 59 Issue number: 12 Page number: 124005 Subjects: Unified Gauge Models; Phase Diagrams; Unified-Field Theories; Supersymmetry; Yang-Mills Theory; Black Holes; Thermodynamics

Martocchia, A.; Karas, V.; Matt, G. Effects of Kerr space-time on spectral features from X-ray illuminated accretion discs Journal Monthly Notices of the Royal Astronomical Society Date: Mar 01 2000 Abstract: The authors performed detailed calculations of the relativistic effects acting on both the reflection continuum and the iron line from accretion discs around rotating black holes. Fully relativistic transfer of both illuminating and reprocessed photons has been considered in Kerr space-time. The authors calculated overall spectra, line profiles and integral quantities, and present their dependences on the black hole angular momentum.

Mathur, Samir D. Resolving the Black Hole Information Paradox Journal International Journal of Modern Physics A Date: Dec 10 2000 Abstract: The recent progress in string theory strongly suggests that formationand evaporation of black holes is a unitary process. This fact makesit imperative that the authors find a flaw in the semiclassical reasoning that implies a loss of information. The authors propose a new criterion that limits the domain of classical gravity & colon; the hyper surfaces of a foliation cannot be stretched too much. This conjectured criterion may have important consequences for the early universe. Journal volume: 15 Issue number: 30 Rights: World Scientific Publishing Company Publisher: World Scientific Publishing Company

Matyjasek, J.; Zaslavskii, O. B. Quantum back reaction of massive fields and self-consistent semiclassical extreme black holes and acceleration horizons Journal Physical Review Date: Nov 15 2001 Abstract: Abstract available at publisher web site Journal volume: 64 Issue number: 10 Page number: 104018-104018-11 Publisher: American Physical Society

Mavromatos, Nick E.; Winstanley, Elizabeth Infinitely coloured black holes Journal Classical and Quantum Gravity Date: Apr 07 2000 Abstract: The authors formulate the field equations for SU() Einstein-Yang-Mills theory, and use an analytic approximation to elucidate the properties of spherically symmetric black hole solutions. This model may be motivated by string theory considerations, given the enormous gauge symmetries which characterize string theory. The solutions simplify considerably in the presence of a negative cosmological constant, particularly for the limiting cases of a very large cosmological constant or very small gauge field. The black holes possess infinite amounts of gauge field hair, and the authors speculate on possible consequences of this for quantum decoherence, which, however, the authors do not tackle here. Journal volume: 17 Issue number: 7 Page number: 1595-1611 Publisher: Institute of Physics

Mayo, Avraham E. Optimal entropy bound and the self-energy of test objects in the vicinity of a black hole Journal Physical Review Date: Nov 15 1999 Abstract: Abstract available at publisher web site Journal volume: 60 Issue number: 10 Page number: 104044-104044-9 Publisher: American Physical Society

Mbonye, Manasse R. Horizon dynamics of evaporating black holes in a higher dimensional inflationary universe Journal Physical Review Date: Dec 15 1999 Abstract: Abstract available at publisher web site Journal volume: 60 Issue number: 12 Page number: 124007-124007-7 Publisher: American Physical Society

McConnell, M; Ryan, J.; Fletcher, S.; Bennett, K; van Dijk, R.; Bloemen, H; Hermsen, W.; Diehl, R; Schoenfelder, V; Strong, A. COMPTEL all-sky imaging at 2.2 MeV Journal AIP Conference Proceedings Date: May 31 1997 Abstract: It is now generally accepted that accretion of matter onto a compact object (white dwarf, neutron star or black hole) is one of the most efficient processes in the universe for producing high energy radiations. Measurements of the {gamma}-ray emission will provide a potentially valuable means for furthering our understanding of the accretion process. Here the authors focus on neutron capture processes, which can be expected in any situation where energetic neutrons may be produced and where the liberated neutrons will interact with matter before they decay (where they have a chance of undergoing some type of neutron capture). Journal volume: 410 Issue number: 1 Page number: 1099-1103 Subjects: Accretion Disks; Cosmic Gamma Sources; Hydrogen; Neutron Stars; Black Holes; White Dwarf Stars; Neutron Reactions

Medved, A. J. M.; Kunstatter, G. One-loop corrected thermodynamics of the extremal and nonextremal spinning Banados-Teitelboim-Zanelli black hole Journal Physical Review Date: May 15 2001 Abstract: Abstract available at publisher web site Journal volume: 63 Issue number: 10 Page number: 104005-104005-12 Publisher: American Physical Society

Medved, A. J. M.; Kunstatter, G. Quantum corrections to the thermodynamics of charged 2D black holes Journal Physical Review Date: Nov 15 1999 Abstract: Abstract available at publisher web site Journal volume: 60 Issue number: 10 Page number: 104029-104029-16 Publisher: American Physical Society

Mena, Filipe C.; Tavakol, Reza; Joshi, Pankaj S. Initial data and spherical dust collapse Journal Physical Review Date: Aug 15 2000 Abstract: Abstract available at publisher web site Journal volume: 62 Issue number: 4 Page number: 044001-044001-8 Publisher: American Physical Society

Merritt, David; Ferrarese, Laura Black hole demographics from the &formmu1; relation Journal Monthly Notices of the Royal Astronomical Society Date: Jan 01 2001 Abstract: The authors analyse a sample of 32 galaxies for which a dynamical estimate of the mass of the hot stellar component, M bulge, is available. For each of these galaxies, the authors calculate the mass of the central black hole, M •, using the tight empirical correlation between M • and bulge stellar velocity dispersion. The authors present marginal evidence for a lower, average black hole mass fraction in more massive galaxies. Journal volume: 320 Issue number: 3 Page number: L30-L34 DOE sponsorship Publisher: Blackwell Science

Michaud, G; Myers, R.C. Hermitian D-brane solutions Journal Physical Review, D (Particles Fields) Date: Sep 30 1997 Abstract: A low-energy background field solution describing D-membrane configurations is constructed which is distinguished by the appearance of a Hermitian metric on the internal space. This metric is composed of a number of independent harmonic functions on the transverse space. Journal volume: 56 Issue number: 6 Page number: 3698-3705 Subjects: Solitons/Hermitian Operators; Solitons/Superstring Models; Solitons; Harmonics; Supersymmetry; Duality; Phase Space; Supergravity; Equations of State; Black Holes; Entropy

Michelson, J. Scattering of four-dimensional black holes Journal Physical Review, D (Particles Fields) Date: Jan 31 1998 Abstract: The moduli space metric for an arbitrary number of extremal black holes in four dimensions with arbitrary relatively supersymmetric charges is found. Journal volume: 57 Issue number: 2 Page number: 1092-1097 Subjects: Four-Dimensional Calculations; Supersymmetry; Duality

Michelson, J. Scattering of four-dimensional black holes Journal Physical Review, Date: Jan 31 1998 Abstract: The moduli space metric for an arbitrary number of extremal black holes in four dimensions with arbitrary relatively supersymmetric charges is found. Journal volume: 572 Page number: 1092-1097

Mignemi, S. Black hole solutions of two-dimensional Riemann–Cartan gravity with higher-derivative action Journal Annals of Physics (New York) Date: Jun 30 1997 Abstract: The authors investigate two-dimensional higher-derivative gravitational theories in a Riemann-Cartan framework and show their equivalence with a first-order formulation with non-trivial potential. The authors obtain the most general solutions in the case of an action quadratic in torsion and discuss their geometry. Journal volume: 257 Issue number: 1 Page number: 1-17 Subjects: Hamiltonian Function; Torsion; General Relativity Theory; Action Integral; Poincare Groups; Quantization; Differential Geometry; Space-Time

Mignemi, S. Black Holes and Conformal Mechanics Journal Modern Physics Letters A Date: Oct 01 2001 Abstract: The authors show how the motion of a charged particle near the horizon of an extreme Reissner-Nordström black hole can lead to different forms of conformal mechanics, depending on the choice of the time coordinate. Journal volume: 16 Issue number: 31 Rights: World Scientific Publishing Company Publisher: World Scientific Publishing Company

Mitra, P. Charged Black Holes in Asymptotically Anti-deSitter Spacetime Journal International Journal of Modern Physics B Date: Aug 10 2000 Abstract: Unlike the extremal Reissner-Nordström black hole in ordinary spacetime, the one in anti-de Sitter spacetime is a minimum of action and has zero entropy if (functional integral) quantization is carried out after extremalization. However, if extremalization is carried out after quantization, then the entropy is a quarter of the area as in the usual case. This is because the functional integral is dominated by non-extremal configurations of lower action. However, some special non-extremal black holes have higher action and may decay into extremal configurations. An attempt is also made to see if the anti-de Sitter version can be described by a microscopic model consisting of a one-dimensional gas of massless particles as in the usual case. Such a description is possible only in the limit of vanishing cosmological constant. Journal volume: 14 Issue number: 19_20 Rights: World Scientific Publishing Company Publisher: World Scientific Publishing Company

Mitra, P. Note on the entropy of charged multi-black-holes Journal Physical Review Date: Jan 15 2000 Abstract: Abstract available at publisher web site Journal volume: 61 Issue number: 2 Page number: 024028-024028-4 Publisher: American Physical Society

Moderski, Rafal; Rogatko, Marek Question of Abelian-Higgs hair expulsion from extremal dilaton black holes Journal Physical Review Date: Nov 15 1999 Abstract: Abstract available at publisher web site Journal volume: 60 Issue number: 10 Page number: 104040-104040-9 Publisher: American Physical Society

Mohaupt, Thomas Black holes in supergravity and string theory Journal Classical and Quantum Gravity Date: Sep 07 2000 Abstract: P The authors give an elementary introduction to black holes in supergravity and string theory. The focus is on the role of BPS solutions in four- and higher-dimensional supergravity and in string theory. Basic ideas and techniques are explained in detail, including exercises with solutions. Journal volume: 17 Issue number: 17 Page number: 3429-3482 Publisher: Institute of Physics

Morales, R.; Fabian, A.C. Weighing black holes with warm absorbers Journal Monthly Notices of the Royal Astronomical Society Date: Jan 01 2002 Abstract: The authors present a new technique for determining an upper limit for the mass of the black hole in active galactic nuclei showing warm absorption features. The method relies on the balance of radiative and gravitational forces acting on outflowing warm absorber clouds. It has been applied to six objects: five Seyfert 1 galaxies: IC 4329a, MCG-6-30-15, NGC 3516, NGC 4051 and NGC 5548; and one radio-quiet quasar: MR 2251-178. The authors discuss our result in comparison with other methods. The procedure could also be applied to any other radiatively driven optically thin outflow in which the spectral band covering the major absorption is directly observed. Journal volume: 329 Issue number: 1 Page number: 209-220 DOE sponsorship Publisher: Blackwell Science Ltd

Morawetz, Cathleen Synge Variations on conservation laws for the wave equation Journal Bulletin of the American Mathematical Society Date: Jan 21 2000 Abstract: The first part of this paper, presented as an Emmy Noether lecture in connection with the ICM in Berlin in August 1998, gives some examples of using Noether's theorem for conservation laws for Tricomi-like equations and for the wave equation. It is also shown that equations which are semilinear variations of the wave equation can very often be handled similarly. The type of estimate obtained can even be used to get otherwise unobtainable local estimates for regularity. The fourth part is an introduction to the relation of black holes to the wave equation mainly showing the results of D.~Christodoulou. Journal volume: 37 Issue number: 02 Page number: 141-154 Rights: Copyright 2000, American Mathematical Society DOE sponsorship Publisher: American Mathematical Society DOI: 10.1090.S0273-0979-00-00857-0

Murali, Chigurupati; Arras, Phil; Wasserman, Ira Constraints on the mass and abundance of black holes in the Galactic halo: the high-mass limit Journal Monthly Notices of the Royal Astronomical Society Date: Mar 01 2000 Abstract: The authors establish constraints on the mass and abundance of black holes in the Galactic halo by determining their impact on globular clusters, which are conventionally considered to be little evolved. Using detailed Monte Carlo simulations and simple evolutionary models, the authors argue that black holes with masses Mbh(1-3)×106 M can comprise no more than a fraction fbh 0.17 of the total halo density at Galactocentric radius R 8 kpc. This bound arises from requiring stability of the cluster mass function.

Myung, Y. S. Quintessence with a Localized Scalar Field on the Brane Journal Modern Physics Letters A Date: Sep 01 2001 Abstract: The authors study issues of the quintessence in the brane cosmology. The initial bulk space-time consists of two 5D topological anti-de Sitter black hole joined by the brane (moving domain wall). Here the authors do not introduce any conventional radiation and matter. Instead the authors include a localized scalar on the brane as a stress-energy tensor, and thus the authors find the quintessence which gives an accelerating universe.

Natsuume, Makoto; Okamura, Takashi Entropy for asymptotically AdS 3 black holes Journal Physical Review Date: Sep 15 2000 Abstract: Abstract available at publisher web site Journal volume: 62 Issue number: 6 Page number: 064027-064027-8 Publisher: American Physical Society

Nayak, K. Rajesh; MacCallum, M. A. H.; Vishveshwara, C. V. Black holes in nonflat backgrounds: The Schwarzschild black hole in the Einstein universe Journal Physical Review Date: Jan 15 2001 Abstract: Abstract available at publisher web site Journal volume: 63 Issue number: 2 Page number: 024020-024020-5 Publisher: American Physical Society

New, Kimberly.C.B.; Shapiro, Stuart L. The formation of supermassive black holes and the evolution of supermassive stars Journal Classical and Quantum Gravity Date: Oct 07 2001 Abstract: P The existence of supermassive black holes is supported by a growing body of observations. Supermassive black holes and their formation events are likely candidates for detection by proposed long-wavelength, space-based gravitational wave interferometers like LISA. However, the nature of the progenitors of supermassive black holes is rather uncertain. Supermassive black hole formation scenarios that involve either the stellar dynamical evolution of dense clusters or the hydrodynamical evolution of supermassive stars have been proposed. Each of these formation scenarios is reviewed and the evolution of supermassive stars is then examined in some detail.

Ng, Y. Jack From Computation to Black Holes and Space-Time Foam Journal Physical Review Letters Date: Apr 02 2001 Abstract: The authors show that quantum mechanics and general relativity limit the speed $\{nu\}$ of a simple computer (such as a black hole) and its memory space I to $I\{nu\}\{sup\ 2\}\{approx\} < t\{sup\ -2\}\{sub\ P\}$, where $t\{sub\ P\}$ is the Planck time. The authors also show that the lifetime of a simple clock and its precision are similarly limited.

Ng, Y. Jack From Computation to Black Holes and Space-Time Foam Journal Physical Review Letters Date: Apr 02 2001 Abstract: Abstract available at publisher web site Journal volume: 86 Issue number: 14 Page number: 2946-2949 Publisher: American Physical Society

Niemeyer, Jens C.; Bousso, Raphael Nonlinear evolution of de Sitter space instabilities Journal Physical Review Date: Jul 15 2000 Abstract: Abstract available at publisher web site Journal volume: 62 Issue number: 2 Page number: 023503-023503-7 Publisher: American Physical Society

Nitta, Shin-ya An application of the Kerr black hole fly-wheel model to statistical properties of QSOs/AGNs Journal Monthly Notices of the Royal Astronomical Society Date: Oct 01 1999 Abstract: The aim of this work is to demonstrate the properties of the magnetospheric model around Kerr black holes (BHs), the so-called fly-wheel (rotation driven) model. The fly-wheel engine of the BH-accretion disc system is applied to the statistics of QSOs/AGNs. In the model, the central BH is assumed to be formed at $z\sim 102$ and obtains nearly maximum but finite rotation energy (~extreme Kerr BH) at the formation stage.

Nojiri, Shin'ichi; Obregon, Octavio; Odintsov, Sergei D.; Ogushi, Sachiko Dilatonic brane-world black holes, gravity localization, and Newton's constant Journal Physical Review Date: Sep 15 2000 Abstract: Abstract available at publisher web site Journal volume: 62 Issue number: 6 Page number: 064017-064017-10 Publisher: American Physical Society

Nojiri, Shinichi; Obregon, Octavio; Odintsov, Sergei D.; Tkach, Vladimir I. String versus Einstein frame in an AdS/CFT induced quantum dilatonic brane-world universe Journal Physical Review Date: Aug 15 2001 Abstract: An AdS/CFT induced quantum dilatonic brane world where the 4D boundary is flat or a de Sitter (inflationary) or anti--de Sitter brane is considered. The classical brane tension is fixed but the boundary QFT produces the effective brane tension by means of the account of corresponding conformal anomaly induced effective action. This results in inducing brane worlds in accordance with the AdS/CFT setup as warped compactification. The explicit, independent construction of quantum induced dilatonic brane worlds in two frames, the string and Einstein frames, is done.

Nojiri, Shin'ichi; Odintsov, Sergei D. Thermodynamics of Schwarzschild (Anti-) de Sitter Black Holes with Account of Quantum Corrections Journal International Journal of Modern Physics A Date: Mar 20 2000 Abstract: The authors discuss the quantum corrections to thermodynamics (and geometry) of S(A)dS BH's using large N one-loop anomaly induced effective action for dilaton coupled matter (scalars and spinors). It is found that the temperature, mass and entropy with account of quantum effects for multiply horizon SdS BH and SAdS BH which also gives the corresponding expressions for their limits & colon; Schwarzschild and de Sitter spaces. In the last case one can talk about quantum correction to the entropy of an expanding universe. The anomaly induced action under discussion corresponds to 4d formulation (s-wave approximation, 4d quantum matter is a minimal one) as well as 2d formulation (complete effective action, 2d quantum matter is a dilaton coupled one).

Nollert, Hans-Peter Quasinormal modes: the characteristic 'sound' of black holes and neutron stars Journal Classical and Quantum Gravity Date: Dec 01 1999 Abstract: Gravitational waves emitted by perturbed black holes or relativistic stars are dominated by 'quasinormal ringing', damped oscillations at single frequencies which are characteristic of the underlying system. These quasinormal modes have been studied for a long time, often with the intent of describing the time evolution of a perturbation in terms of these modes in a way very similar to a normal-mode analysis. In this review, the authors summarize how quasinormal modes are defined and computed.

Nucamendi, Ulises; Sudarsky, Daniel Black holes with zero mass Journal Classical and Quantum Gravity Date: Oct 07 2000 Abstract: P The authors consider the spacetimes corresponding to static global monopoles with interior boundaries corresponding to a black hole horizon and analyse the behaviour of the appropriate ADM mass as a function of the horizon radius $r H$. The authors find that for small enough $r H$, this mass is negative as in the case of the regular global monopoles, but that for large enough $r H$ the mass becomes positive, encountering an intermediate value for which the authors have a black hole with zero ADM mass. Journal volume: 17 Issue number: 19 Page number: 4051-4058 Publisher: Institute of Physics

Olum, K.D. Entropy of localized states and black hole evaporation Journal Physical Review, D (Particles Fields) Date: May 31 1997 Abstract: The authors call a state "vacuum bounded" if every measurement performed outside a specified interior region gives the same result as in the vacuum. The authors compute the maximum entropy of a vacuum-bounded state with a given energy for a one-dimensional model, with the aid of numerical calculations on a lattice. Journal volume: 55 Issue number: 10 Page number: 6168-6180 Subjects: Black Holes/Evaporation; Quantum Gravity; Evaporation; Vacuum States; Locality; Entropy; Boundary Conditions

Olum, K.D. Entropy of very low energy localized states Journal Physical Review, Date: Feb 28 1998 Abstract: The authors expand on previous work involving "vacuum-bounded" states, i.e., states such that every measurement performed outside a specified interior region gives the same result as in the vacuum. The authors improve our previous techniques by removing the need for a finite outside region in numerical calculations. The authors apply these techniques to the limit of very low energies and show that the entropy of a vacuum-bounded state can be much higher than that of a rigid box state with the same energy. Journal volume: 574 Page number: 2486-2499 Subject keyword: Black Holes; Bound State; Density Matrix; Energy; Entropy; Expectation Value; Locality; Numerical Analysis; Vacuum States; Mathematics; Matrices; Physical Properties; Thermodynamic Properties

Onozawa, H; Okamura, T.; Mishima, T.; Ishihara, H. Perturbing supersymmetric black holes Journal Physical Review, D (Particles Fields) Date: Apr 30 1997 Abstract: An investigation of the perturbations of the Reissner-Nordström black hole in $N=2$ supergravity is presented. The authors prove in the extremal limit that the black hole responds to the perturbation of each field in the same manner. Journal volume: 55 Issue number: 8 Page number: R4529-R4531 Subjects: Black Holes/supersymmetry; Supersymmetry; Perturbation Theory; Supergravity; Gravitons; Transformations; Photons; Sparticles

Opher, Reuven Introduction to plasma astrophysics Journal Plasma Physics and Controlled Fusion Date: Mar 01 1999 Abstract: The authors discuss the importance of plasma physics in various areas of cosmology and astrophysics: (1) the primordial plasma spectrum, (2) the cosmic microwave background, (3) the large-scale structure of the universe, (4) reionization and the metalicity of the primordial universe, (5) the first objects, (6) the primordial magnetic field, (7) clusters of galaxies, (8) galaxy formation, (9) star formation, (10) nuclear reactions, (11) accretion onto black holes, (12) cosmic rays, (13) relativistic jets and (14) gamma ray bursters. Journal volume: 41 Issue number: 3A Page number: A209-A220 Publisher: Institute of Physics

Ori, Amos Oscillatory Null Singularity inside Realistic Spinning Black Holes Journal Physical Review Letters Date: Dec 27 1999 Abstract: Abstract available at publisher web site Journal volume: 83 Issue number: 26 Page number: 5423-5426 Publisher: American Physical Society

Page, Don N. Black holes with less entropy than A/4 Journal Physical Review Date: Jan 15 2002 Abstract: Abstract available at publisher web site Journal volume: 65 Issue number: 2 Page number: 024017-024017-12 Publisher: American Physical Society

Page, M. J.; Stevens, J. A.; Mittaz, J. P. D.; Carrera, F. J. Submillimeter Evidence for the Coeval Growth of Massive Black Holes and Galaxy Bulges Journal Title: Science Magazine Date: Dec 21 2001 Abstract:

Abstract available at publisher web site Journal volume: 294 Issue number: 5551 Page number: 2516-2518 DOE sponsorship Publisher: American Association for the Advancement of Science

Papadopoulos, Philippos; Font, José A. Imprints of accretion on gravitational waves from black holes Journal Physical Review Date: Feb 15 2001 Abstract: Abstract available at publisher web site Journal volume: 63 Issue number: 4 Page number: 044016-044016-5 Publisher: American Physical Society

Pariev, V.I.; Bromley, B.C. Line Emission from an Accretion Disk Around a Black Hole: Effects of Disk Structure Journal Astrophysical Journal Date: Dec 31 1998 Abstract: The observed iron K{alpha} fluorescence lines in Seyfert I galaxies provide strong evidence for an accretion disk near a supermassive black hole as a source of the line emission. These lines serve as powerful probes for examining the structure of inner regions of accretion disks. Previous studies of line emission have considered only geometrically thin disks, where the gas moves along geodesics in the equatorial plane of a black hole. Here the authors extend this work to consider the effects on line profiles from finite disk thickness, radial accretion flow, and turbulence. Journal volume: 508 Issue number: 2 Page number: 590-600 Subjects: Accretion Disks; Fluid Mechanics; Seyfert Galaxies; Spectra; Line Widths

Parikh, M.K.; Wilczek, F. An action for black hole membranes Journal Physical Review, D (Particles Fields) Date: Sep 30 1998 Abstract: The membrane paradigm is the remarkable view that, to an external observer, a black hole appears to behave exactly like a dynamical fluid membrane, obeying such pre-relativistic equations as Ohm's law and the Navier-Stokes equation. It has traditionally been derived by manipulating the equations of motion. Here the authors provide an action formulation of this picture, clarifying what underlies the paradigm and simplifying the derivations. Journal volume: 58 Issue number: 6 Page number: 064011 Subjects: Einstein-Maxwell Equations; Black Holes; Membranes; Fluids; Ohm Law; Navier-Stokes Equations; Equations Of Motion; Feynman Path Integral; Hamiltonians; Entropy; Thermodynamics; Dyons

Park, Mu-In; Yee, Jae Hyung Comment on "Entropy of 2D black holes from counting microstates" Journal Physical Review Date: Apr 15 2000 Abstract: Abstract available at publisher web site Journal volume: 61 Issue number: 8 Page number: 088501-088501-2 Publisher: American Physical Society

Paul, B. C. Probability for primordial black holes in a higher dimensional universe Journal Physical Review Date: Jan 15 2000 Abstract: Abstract available at publisher web site Journal volume: 61 Issue number: 2 Page number: 024032-024032-4 Publisher: American Physical Society

Pavlidou, Vasiliki; Tassis, Konstantinos; Baumgarte, Thomas W.; Shapiro, Stuart L. Radiative falloff in neutron star spacetimes Journal Physical Review Date: Oct 15 2000 Abstract: Abstract available at publisher web site Journal volume: 62 Issue number: 8 Page number: 084020-084020-8 American Physical Society

Peet, Amanda W. The Bekenstein formula and string theory (N-brane theory) Journal Classical and Quantum Gravity Date: Nov 01 1998 Abstract: A review of recent progress in string theory concerning the Bekenstein formula for black hole entropy is given. Topics discussed include p-branes, D-branes and supersymmetry; the correspondence principle; the D- and M-brane approach to black hole entropy; the D-brane analogue of Hawking radiation and information loss; D-branes as probes of black holes and the matrix theory approach to charged and neutral black holes. Some introductory material is included, as are some very brief remarks on the AdS/CFT correspondence. Journal volume: 15 Issue number: 11 Page number: 3291-3338 Publisher: Institute of Physics

Peña, Igor; Sudarsky, Daniel Do collapsed boson stars result in new types of black holes? Journal Classical and Quantum Gravity Date: Nov 01 1997 Abstract: The authors prove a general no-hair theorem which is then applied to the case of a theory that consists of a number of complex scalar fields minimally coupled to gravity that vary harmonically with time and has an arbitrary potential. This establishes that in the spherically symmetric case there are no black hole analogues of the regular boson star configurations. Journal volume: 14 Issue number: 11 Page number: 3131-3134 Publisher: Institute of Physics

Peña, Igor; Sudarsky, Daniel Numerical evidence against black holes with non-minimally coupled scalar hair Journal Classical and Quantum Gravity Date: Apr 21 2001 Abstract: P The authors search numerically for asymptotically flat, static and spherically symmetric black hole configurations in a system which consists of a real scalar field non-minimally coupled to gravity. This issue has been analysed previously, assuming external considerations to the theory which, as the authors explain, limit the generality of the analysis. The authors present a method that is useful for obtaining initial conditions for numerical analysis. Our results narrow the range in which one can find black hole configurations without considering any requirements external to the theory. Journal volume: 18 Issue number: 8 Page number: 1461-1474 Publisher: Institute of Physics

Peterson, Bradley M.; Wandel, Amri Evidence for Supermassive Black Holes in Active Galactic Nuclei from Emission-Line Reverberation Journal The Astrophysical Journal Date: Aug 01 2000 Abstract: Emission-line variability data for Seyfert 1 galaxies provide strong evidence for the existence of supermassive black holes in the nuclei of these galaxies and that the line-emitting gas is moving in the gravitational potential of that black hole. The time-delayed response of the emission lines to continuum variations is used to infer the size of the line-emitting region, which is then combined with measurements of the Doppler widths of the variable line components to estimate a virial mass.

Petrucci, P.O.; Merloni, A.; Fabian, A.; Haardt, F.; Gallo, E. The effects of a Comptonizing corona on the appearance of the reflection components in accreting black hole spectra Journal Monthly Notices of the Royal Astronomical Society Date: Dec 01 2001 Abstract: The authors discuss the effects of a Comptonizing corona on the appearance of the reflection components, and in particular of the reflection hump, in the X-ray spectra of accreting black holes. Indeed, in the framework of a thermal corona model, the authors expect that some (or even all, depending on the coronal covering factor) of the reflection features should cross the hot plasma, and thus suffer Compton scattering, before being observed.

Pfeiffer, Harald P.; Teukolsky, Saul A.; Cook, Gregory B. Quasicircular orbits for spinning binary black holes Journal Physical Review Date: Nov 15 2000 Abstract: Abstract available at publisher web site Journal volume: 62 Issue number: 10 Page number: 104018-104018-11 Publisher: American Physical Society

Phillips, Nicholas G.; Hu, B. L. Noise kernel in stochastic gravity and stress energy bitensor of quantum fields in curved spacetimes Journal Physical Review Date: May 15 2001 Abstract: Abstract available at publisher web site Journal volume: 63 Issue number: 10 Page number: 104001-104001-16 Publisher: American Physical Society

Pierre, J.M. Comparing D-branes and black holes with 0- and 6-brane charges Journal Physical Review, D (Particles Fields) Date: Nov 30 1997 Abstract: The authors consider configurations of D6-branes with a D0-brane charge given by recent work of Taylor and compute interaction potentials with various D-brane probes using a 1-loop open string calculation. These results are compared to a supergravity calculation using the solution given by Sheinblatt of an extremal black hole carrying 0-brane and 6-brane charges. Journal volume: 56 Issue number: 10 Page number: 6710-6713 Subjects: Charges; String Models; Supergravity; Solitons; Space-Time

Plyatsko, Roman; Bilaniuk, Oleksa Gravitational ultrarelativistic interaction of classical particles in the context of unification of interactions Journal Classical and Quantum Gravity Date: Dec 07 2001 Abstract: The response of the ultrarelativistic particle with spin in a Schwarzschild field to the gravitomagnetic components as measured by the co-moving observer is investigated. The dependence of the particle's spin-orbit acceleration on the Lorentz γ-factor and the spin orientation is studied. The concrete circular ultrarelativistic orbit of radius r &equal; 3m is considered as a partial solution of the Mathisson-Papapetrou equations and as the corresponding high-energy quantum state of the Dirac particle.

Podolsky´, J.; Griffiths, J. B. Uniformly accelerating black holes in a de Sitter universe Journal Physical Review Date: Jan 15 2001 Abstract: Abstract available at publisher web site Journal volume: 63 Issue number: 2 Page number: 024006-024006-6 Publisher: American Physical Society

Polarski, D.; Dolgov, A. Classicality of Primordial Fluctuations and Primordial Black Holes Journal International Journal of Modern Physics A Date: Dec 01 2001 Abstract: The production of Primordial Black Holes (PBH) from inflationary perturbations provides a physical process where the effective classicality of the fluctuations does not hold for certain scales. For adiabatic perturbations produced during inflation, this range of scales corresponds to PBH with masses M≪1015 g. For PBH with masses M ~ M H (te), the horizon mass at the end of inflation, the generation process during the preheating stage could be classical as well, in contrast to the formation of PBH on these scales by adiabatic inflationary perturbations. For then on evaporated PBH, the generation process is essentially classical. Journal volume: 10 Issue number: 06 Rights: World Scientific Publishing Company Publisher: World Scientific Publishing Company

Pons, Jose A.; Steiner, Andrew W.; Prakash, Madappa; Lattimer, James M. Evolution of Proto-Neutron Stars with Quarks Journal Physical Review Letters Date: Jun 04 2001 Abstract: Neutrino fluxes from proto-neutron stars with and without quarks are studied. Observable differences become apparent after 10–20s of evolution. Sufficiently massive stars containing negatively charged, strongly interacting, particles collapse to black holes during the first minute of evolution. Since the neutrino flux vanishes when a black hole forms, this is the most obvious signal that quarks (or other types of strange matter) have appeared. Journal volume: 86 Issue number: 23 Page number: 5223-5226 Publisher: American Physical Society

Portegies Zwart, Simon F.; McMillan, Stephen L. W. Gravitational Thermodynamics and Black-Hole Mergers Journal International Journal of Modern Physics A Date: Dec 10 2000 Abstract: Black holes become the most massive objects early in the evolution of star clusters. Dynamical relaxation then causes them to sink to the cluster core, where they form binaries which become more tightly bound by superelastic encounters with other cluster members. Ultimately, these binaries are ejected from the cluster. The majority of escaping black-hole binaries have orbital periods short enough and eccentricities high enough that the emission of gravitational waves causes them to coalesce within a few billion years. The rate at which such collisions occur is on the order of 10-7 per year per cubic megaparsec.

Pravda, V.; Pravdová, A. Co-accelerated particles in the C-metric Journal Classical and Quantum Gravity Date: Apr 07 2001 Abstract: P With appropriately chosen parameters, the C-metric represents two uniformly accelerated black holes moving in the opposite directions on the axis of the axial symmetry (the z-axis). The acceleration is caused by nodal singularities located on the z-axis. /PP In the present paper, geodesics in the C-metric are examined. In general, there exist three types of timelike or null geodesics in the C-metric: geodesics describing particles (a) falling under the black hole horizon; (b) crossing the acceleration horizon; and (c) orbiting around the z-axis and co-accelerating with the black holes.

Prestidge, Tim Dynamic and thermodynamic stability and negative modes in Schwarzschild-anti-de Sitter black holes Journal Physical Review Date: Apr 15 2000 Abstract: Abstract available at publisher web site Journal volume: 61 Issue number: 8 Page number: 084002-084002-13 Publisher: American Physical Society

Pretorius, Frans; Choptuik, Matthew W. Gravitational collapse in 2+1 dimensional AdS spacetime Journal Physical Review Date: Dec 15 2000 Abstract: Abstract available at publisher web site Journal volume: 62 Issue number: 12 Page number: 124012-124012-15 Publisher: American Physical Society

Price, Richard H.; Whelan, John T. Tidal Interaction in Binary-Black-Hole Inspiral Journal Physical Review Letters Date: Dec 03 2001 Abstract: Abstract available at publisher web site Journal volume: 87 Issue number: 23 Page number: 231101-231101-4 American Physical Society

Qian, Y; Vogel, P.; Wasserburg, G.J. Diverse supernova sources for the r-process Journal Astrophysical Journal Date: Feb 28 1998 Abstract: The authors present a simplified analysis using equations for the charge flow, which include nu_e capture, for the production of r-process nuclei in the context of the recent supernova hot bubble model. Journal volume: 494 Issue number: 1 Page number: 285-296 Subjects: Supernovae; Mathematical Models; Nucleosynthesis; Element Abundance

Qiu, Weigang; Wang, Bin; Su, Ru-Keng; Abdalla, Elcio Entropy bound for a charged rotating system Journal Physical Review Date: Jul 15 2001 Abstract: Abstract available at publisher web site Journal volume: 64 Issue number: 2 Page number: 027503-027503-4 Publisher: American Physical Society

Qiu, Weigang; Xu, Jianjun; Su, Ru-Keng; Wang, Bin Quantum Corrections to the Entropy of Extreme Reissner-Nordström Black Holes Journal Modern Physics Letters A Date: Mar 01 2001 Abstract: Using brick wall model, the first quantum corrections to the extreme Reissner-Nordström black hole entropy due to scalar field as well as electromagnetic field have been calculated. Different quantum entropy values have been obtained for two different kinds of extreme black holes. Journal volume: 16 Issue number: 08 Publisher: World Scientific Publishing Company

Ramon Medrano, M.; Sanchez, N. Hawking radiation in string theory and the string phase of black holes Journal Physical Review Date: Apr 15 2000 Abstract: Abstract available at publisher web site Journal volume: 61 Issue number: 8 Page number: 084030-084030-11 Publisher: American Physical Society

Rees, Martin J. Gravitational waves from galactic centres? Journal Classical and Quantum Gravity Date: Jun 01 1997 Abstract: Young galaxies probably acquire central black holes which (during and soon after their formation) manifest themselves as quasars and may be reactivated later. It seems unlikely that the holes form suddenly enough to be intense sources of gravitational radiation. Their formation and growth may, however, yield lower level emission. There is a guarantee of intense bursts from later mergers of supermassive holes and of weak nearly periodic waves from stars captured into relativistic orbits around massive holes. Journal volume: 14 Issue number: 6 Page number: 1411-1415 Publisher: Institute of Physics

Reid, D.D. Discrete quantum gravity and causal sets Journal Canadian Journal of Physics Date: Jan 01 2001 Abstract: This paper provides a thorough introduction to the physical and conceptual need for a theory of quantum gravity; some knowledge of general relativity and nonrelativistic quantum mechanics is assumed. A theory of quantum gravity would have wide-ranging implications for high-energy physics, astrophysics, and cosmology. The paper goes on to describe an important approach to quantum gravity that is not well known outside of the quantum gravity research community — causal sets. The causal-set approach falls within the framework of discrete quantum gravity, which considers the possibility that the small-scale structure of spacetime might be discrete rather than continuous.

Reig, P.; D., N.; C., H. Orbital Comptonization in accretion disks around black holes Journal Astronomy and Astrophysics Date: Jun 07 2001 Abstract: The authors have performed Monte Carlo simulations of Compton upscattering of low-energy photons in an accretion disk around a Schwarzschild blackhole. The photons gain energy from the rotational motion of the electrons in the disk. The upscattering occurs near the black hole horizon, where the flow velocity of the electrons approaches the speed of light. The authors show that this type of bulk-flow Comptonization can produce power-law X-ray spectra similar to the ones observed in black-hole X-ray transients in the high/soft state, i.e., a soft bump dominating the spectrum below ~10 ke V and a power-law tail with photon index in the range 2-3.

Rembges, F; Freiburghaus, C; Rauscher, T; Thielemann, F.; Schatz, H; Wiescher, M. An approximation for the rp-process Journal Astrophysical Journal Date: Jul 31 1997 Abstract: Hot (explosive) hydrogen burning, or the rapid proton capture process (rp-process), occurs in a number of astrophysical environments. Novae and X-ray bursts are the most prominent ones, but accretion disks around black holes and other sites are candidates as well. The expensive and often multidimensional hydro calculations for such events require an accurate prediction of the thermonuclear energy generation while avoiding full nucleo synthesis network calculations. Journal volume: 484 Issue number: 1 Page number: 412-423 Subjects: Variable Stars; Hydrogen Burning; Element Abundance; X-Ray Spectra

Revnivtsev, Mikhail G.; Borozdin, Konstantin N.; Priedhorsky, William C.; Vikhlinin, Alexey Rossi X-Ray Timing Explorer Observations of an Outburst of Recurrent X-Ray Nova GS 1354-644 Journal Astrophysical Journal Date: Feb 20 2000 Abstract: The authors present the results of Rossi X-Ray Timing Explorer (RXTE) observations of GS 1354-644 during a modest outburst in 1997-1998. The source is one of a handful of black hole X-ray transients that are confirmed to be recurrent in X-rays. A 1987 outburst of the same source observed by Ginga was much brighter and showed a high/soft spectral state. In contrast, the 1997-1998 outburst showed a low/hard spectral state. Both states are typical of black hole binaries.

Reznik, B. Unruh effect with back reaction A first-quantized treatment Journal Physical Review, D (Particles Fields) Date: Feb 28 1998 Abstract: The authors present a first-quantized treatment of the back reaction on an accelerated particle detector. The evaluated transition amplitude for detection agrees with previously obtained results. Journal volume: 57 Issue number: 4 Page number: 2403-2409 Subjects: Quantum Gravity; Quantization; Transition Amplitudes; Backscattering; Radiations

Reznik, B. Unruh effect with back reaction[e#151]A first-quantized treatment Journal Physical Review, Date: Feb 28 1998 Abstract: The authors present a first-quantized treatment of the back reaction on an accelerated particle detector. The evaluated transition amplitude for detection agrees with previously obtained results. Journal volume: 574 Page number: 2403-2409 Subject keyword: Acceleration; Backscattering; Black Holes; Quantization;

Rogatko, Marek Dilaton black holes on thick branes Journal Physical Review Date: Sep 15 2001 Abstract: Abstract available at publisher web site Journal volume: 64 Issue number: 6 Page number: 064014-064014-8 American Physical Society

Rogatko, Marek Positive mass theorem for black holes in Einstein-Maxwell axion-dilaton gravity Journal Classical and Quantum Gravity Date: Jan 07 2000 Abstract: The authors present a proof of the positive mass theorem for black holes in Einstein-Maxwell axion-dilaton gravity, which is the low-energy limit of heterotic string theory. The authors show that the total mass of a spacetime containing a black hole is greater than or equal to the square root of the sum of the squares of the adequate dilaton-electric and dilaton-magnetic charges. Journal volume: 17 Issue number: 1 Page number: 11-17 Publisher: Institute of Physics

Rogatko, Marek The staticity problem for non-rotating black holes in Einstein - Maxwell axion - dilaton gravity Journal Classical and Quantum Gravity Date: Aug 01 1997 Abstract: Conditions for staticity are investigated in the case of non-rotating black holes in Einstein - Maxwell axion - dilaton gravity. It is proved that given a vanishing asymptotic value of the quantity assembled by means of -duals to the U(1) gauge fields and fulfilling the additional inequality, the metric staticity condition and the staticity condition for the field in the theory under consideration will be satisfied in the domain of outer communication. Journal volume: 14 Issue number: 8 Page number: 2425-2434 Publisher: Institute of Physics

S.A. Hayward Gravitational waves, black holes and cosmic strings in cylindrical symmetry Journal Classical and Quantum Gravity Date: Oct 07 2000 Abstract: Equations (50)-(51) should read/PP ds 2=r -4mdz 2 +r 4m+2 dϕ2+ (1-8 ϵ)-1 e8m(1+4m) r4m(1+2m) (dr 2-dt 2) (50) /PPE=1/8-(1/8-ϵ)e-8m(1+4m) r-8m 2 (51) /PP The author thanks Bill Bonnor for pointing out the problem with the original form of the metric. Journal volume: 17 Issue number: 19 Page number: 4159-4159 Publisher: Institute of Physics

S.P., de Alwis,; Sato, K. Radiation from a class of string theoretic black holes Journal Physical Review, Date: May 31 1997 Abstract: The emission of a scalar with low energy [omega], from a D- (4[le]D[le]8) dimensional black hole with n charges, is studied in both string and semiclassical calculations. In the lowest order in [omega], the weak coupling string and semiclassical calculations agree provided that the Bekenstein-Hawking formula is valid and the effective central charge c[sub eff]=6 for any D. Journal volume: 5510 Page number: 6181-6188 Subjects: Black Holes-- Radiations; Black Holes-- String Models; Charges; Coupling; Entropy; Semiclassical Approximation; Composite Models; Extended Particle Model; Mathematical Models; Particle Models; Physical Properties; Quark Model; Thermodynamic Properties

Sachs, Ivo; Solodukhin, Sergey N. Horizon holography Journal Physical Review Date: Dec 15 2001 Abstract: Abstract available at publisher web site Journal volume: 64 Issue number: 12 Page number: 124023-124023-11 Publisher: American Physical Society

Sakai, Nobuyuki; Barrow, John D. Cosmological evolution of black holes in Brans-Dicke gravity Journal Classical and Quantum Gravity Date: Nov 21 2001 Abstract: The authors consider a modified &lquot; Swiss cheese&rquot; model in the Brans-Dicke theory and use it to discuss the evolution of black holes in an expanding universe. They define the black hole radius by the Misner-Sharp mass and find the exact time evolutions for dust and vacuum universes of all curvatures. Journal volume: 18 Issue number: 22 Page number: 4717-4723 Publisher: Institute of Physics

Quantum Gravity; Radiations; Transition Amplitudes; Amplitudes; Field Theories; Quantum Field Theory; Scattering

Salucci, Paolo; Szuszkiewicz, Ewa; Monaco, Pierluigi; Danese, Luigi Mass function of dormant black holes and the evolution of active galactic nuclei Journal Monthly Notices of the Royal Astronomical Society Date: Aug 01 1999 Abstract: Under the assumption that accretion on to massive black holes (BHs) powers active galactic nuclei (AGNs), the mass function (MF) of the BHs responsible for their past activity is estimated. For this, the authors take into account not only the activity related to the optically selected AGNs, but also that required to produce the hard X-ray background (HXRB). The MF of the massive dark objects (MDOs) in nearby quiescent galaxies is computed by means of the most recent results on their demography.

Sambruna, Rita M.; Chartas, George; Eracleous, Michael; Mushotzky, Richard F.; Nousek, John A. Chandra Uncovers a Hidden Low-Luminosity Active Galactic Nucleus in the Radio Galaxy Hydra A (3C 218) Journal The Astrophysical Journal Date: Apr 01 2000 Abstract: The authors report the detection with Chandra of a low-luminosity active galactic nucleus (LLAGN) in the low-ionization nuclear emission-line region (LINER) hosted by Hydra A, a nearby (z=0.0537) powerful FR I radio galaxy with complex radio and optical morphology. In a 20 ks ACIS-S exposure during the calibration phase of the instrument, a point source is detected at energies 2 keV at the position of the compact radio core, embedded in diffuse thermal X-ray emission (kT~1 keV) at softer energies. The spectrum of the point source is well fitted by a heavily absorbed power law with intrinsic column density.

Santos, Caroline; Gregory, Ruth Vortices and black holes in dilatonic gravity Journal Physical Review Date: Jan 15 2000 Abstract: Abstract available at publisher web site Journal volume: 61 Issue number: 2 Page number: 024006-024006-12 Publisher: American Physical Society

Sarbach, O.; Heusler, M.; Brodbeck, O. Perturbation theory for self-gravitating gauge fields: The odd-parity sector Journal Physical Review Date: Oct 15 2000 Abstract: Abstract available at publisher web site Journal volume: 62 Issue number: 8 Page number: 084001-084001-24 Publisher: American Physical Society

Sarbach, O.; Winstanley, E. On the linear stability of solitons and hairy black holes with a negative cosmological constant: the odd-parity sector Journal Classical and Quantum Gravity Date: Jun 07 2001 Abstract: P Using a recently developed perturbation formalism based on curvature quantities, the authors investigate the linear stability of black holes and solitons with Yang-Mills hair and a negative cosmological constant. The authors show that those solutions which have no linear instabilities under odd- and even-parity spherically symmetric perturbations remain stable under odd-parity, linear, non-spherically symmetric perturbations. Journal volume: 18 Issue number: 11 Page number: 2125-2146 Publisher: Institute of Physics

Sarbach, Olivier; Tiglio, Manuel Gauge-invariant perturbations of Schwarzschild black holes in horizon-penetrating coordinates Journal Physical Review Date: Oct 15 2001 Abstract: Abstract available at publisher web site Journal volume: 64 Issue number: 8 Page number: 084016-084016-15 Publisher: American Physical Society

Sathyaprakash, B.S. Mother templates for gravitational wave chirps Journal Classical and Quantum Gravity Date: Dec 07 2000 Abstract: P Templates used in the search for binary black holes and neutron stars in gravitational wave interferometer data have to be computed on-line since the computational storage and retrieval costs for a template bank would be too expensive. The conventional dimensionless variable $T = (c^3/Gm)t$, where m is the total mass of a binary, in the time domain and a not-so-conventional velocity-like variable $v = (\pi Gmf)^{1/3}$ in the Fourier domain, render the phasing of the waves independent of the total mass of the system enabling the construction of mother templates that depend only on the mass ratio of a black hole binary. Use of such mother templates in a template bank will bring about a reduction in computational costs by up to a factor of 10 and a saving on storage by a factor of 100. Journal volume: 17 Issue number: 23 Page number: L157-L162 Publisher: Institute of Physics

Scardigli, Fabio Black hole entropy: a spacetime foam approach Journal Classical and Quantum Gravity Date: Jul 01 1997 Abstract: I introduce the spacetime foam structure by briefly reviewing the ideas of Wheeler, topological fluctuations and the possibility of virtual black holes. The contribution of Jacobson (the equation of state of the foam) is recalled. P In the second part, I introduce a model of spacetime foam at the surface of the event horizon of a black hole. I apply these ideas to the calculus of the number of states of a black hole, of its entropy and of other thermodynamical properties. A formula for the number of micro-holes on the surface of the event horizon is derived. P Subsequently, I extend the thermodynamical properties of the event horizon to thermodynamical properties of the space. Here I face the problem of the maximum entropy contained in a space region of a given volume. Finally, on the basis of the results obtained previously, I briefly treat the possibility of micro-black-hole creation by the Unruh effect. Journal volume: 14 Issue number: 7 Page number: 1781-1793 Publisher: Institute of Physics

Schneider, R.; Ferrari, V.; Matarrese, S.; Portegies Zwart, S. F. Low-frequency gravitational waves from cosmological compact binaries Journal Monthly Notices of the Royal Astronomical Society Date: Jul 01 2001 Abstract: The authors consider gravitational waves emitted by various populations of compact binaries at cosmological distances. The authors use population synthesis models to characterize the properties of double neutron stars, double black holes and double white dwarf binaries, and white dwarf-neutron star, white dwarf-black hole and black hole-neutron star systems. The authors use the observationally determined cosmic star formation history to reconstruct the redshift distribution of these sources and their merging rate evolution.

Schutz, B.F. Gravitational wave astronomy Journal Classical and Quantum Gravity Date: Dec 01 1999 Abstract: The first decade of the new millennium should see the first direct detections of gravitational waves. This will be a milestone or fundamental physics and it will open the new observational science of gravitational wave astronomy. But gravitational waves already play an important role in the modeling of astrophysical systems. I review here the present state of gravitational radiation theory in relativity and astrophysics, and I then look at the development of detector sensitivity over the next decade, both on the ground (such as LIGO) and in space (LISA). I review the sources of gravitational waves that are likely to play an imp rtant role in observations by first- and second-generation interferometers, including the astrophysical information that will c me from these observations. The review covers some 10 decades of gravitational wave frequency, from the high-frequency normal odes of neutron stars down to the lowest frequencies observable from space. Journal volume: 16 Issue number: 12A Page number: A131-A156 Institute of Physics

Schutz, Bernard Editorial Journal Classical and Quantum Gravity Date: Oct 07 2001 Abstract: PThe Third International LISA Symposium was held at the Max-Planck-Institute for Gravitational Physics (Albert Einstein Institute) in Golm, near Berlin, on 11-14 July 2000. This special issue of Classical and Quantum Gravity contains the proceedings of this meeting./PPLISA is one of the most exciting and challenging scientific space missions ever proposed. It will search for gravitational waves from all parts of the Universe, using three independent spacecraft in a triangular arrangement with separations of 5×106 km. It will open up the low frequency region between 0.1 mHz and 0.1 Hz to direct observation. In this range the authors expect LISA to detect bursts of gravitational radiation lasting several months from mergers between giant supermassive black holes in the centres of galaxies, continuous radiation from thousands of binary star systems in our Galaxy, and much more.

Schützhold, Ralf Particle definition in the presence of black holes Journal Physical Review Date: Jan 15 2001 Abstract: Abstract available at publisher web site Journal volume: 63 Issue number: 2 Page number: 024014-024014-20 Publisher: American Physical Society

Semerák, O.; Bicák, J. The interplay between forces in the Kerr - Newman field Journal Classical and Quantum Gravity Date: Nov 01 1997 Abstract: The authors discuss dynamical properties of generally non-Keplerian equatorial circular orbits and zero-angular-momentum spherical polar orbits around Kerr - Newman black holes. By considering charged test particles, the thrust is simply represented by the Lorentz force due to the electromagnetic field of the hole. The authors analyse the properties of the rotospheres in which the acceleration of the particles depends on their orbital angular velocity in a counter-intuitive manner. Journal volume: 14 Issue number: 11 Page number: 3135-3147 Publisher: Institute of Physics

Semerák, O.; Zácek, M. Gravitating discs around a Schwarzschild black hole: I Journal Classical and Quantum Gravity Date: Apr 07 2000 Abstract: The authors study the class of Lemos-Letelier annular discs (realistic gravitating static axisymmetric thin discs obtained by inversion of the first Morgan-Morgan solution) around Schwarzschild black holes. For each value of the inner radius, the range of masses is found when all of the disc can be interpreted by counter-rotating streams of particles on stable timelike geodesics.

Sermutlu, Emre The first law of black-hole thermodynamics for black holes in string theory Journal Classical and Quantum Gravity Date: Jun 01 1998 Abstract: The authors investigate thermodynamical properties of four- and five-dimensional black-hole solutions of toroidally compactified string theory. The authors find an explicit expression for the first law of black-hole thermodynamics. The authors calculate the temperature T, angular velocity and the electromagnetic potentials on the horizon using two different methods. Journal volume: 15 Issue number: 6 Page number: 1449-1461 Publisher: Institute of Physics

Shi, X; Fuller, G.M.; Halzen, F. Observing the Birth of Supermassive Black Holes with the Planned ICECUBE Neutrino Detector Journal Physical Review Letters Date: Dec 31 1998 Abstract: It has been suggested that the supermassive black holes, at the centers of galaxies and quasars, may initially form in single collapses of relativistic star clusters or supermassive stars built up during the evolution of dense star clusters. The authors show that it may be possible for ICECUBE (a planned 1 thinspthinspkm$\{sup 3\}$ neutrino detector in Antarctica) to detect the neutrino bursts associated with those collapses at redshift z$\{approx_lt\}$0.2 with a rate of $\{approximately\}$ 0.1– 1 thinspthinspburst per year. Journal volume: 81 Issue number: 26 Page number: 5722-5725 Subjects: Quasars; Star Models;Neutrino Detection; Red Shift

Shibata, Masaru; Baumgarte, Thomas W.; Shapiro, Stuart L. Stability and collapse of rapidly rotating, supramassive neutron stars: 3D simulations in general relativity Journal Physical Review Date: Feb 15 2000 Abstract: Abstract available at publisher web site Journal volume: 61 Issue number: 4 Page number: 044012-044012-11 American Physical Society

Shibata, Masaru; Sasaki, Misao Black hole formation in the Friedmann universe: Formulation and computation in numerical relativity Journal Physical Review Date:

Oct 15 1999 Abstract: Abstract available at publisher web site Journal volume: 60 Issue number: 8 Page number: 084002-084002-11 American Physical Society

Shibata, Masaru; Shapiro, Stuart L.; Uryū, Kōji Equilibrium and stability of supermassive stars in binary systems Journal Physical Review Date: Jul 15 2001 Abstract: Abstract available at publisher web site Journal volume: 64 Issue number: 2 Page number: 024004-024004-14 Publisher: American Physical Society

Shinkai, Hisa-aki; Hayward, Sean A. Quasispherical approximation for rotating black holes Journal Physical Review Date: Aug 15 2001 Abstract: The authors numerically implement a quasispherical approximation scheme for computing gravitational waveforms for coalescing black holes, testing it against angular momentum by applying it to Kerr black holes. As error measures, the authors take the conformal strain and specific energy due to spurious gravitational radiation. The strain is found to be monotonic rather than wavelike.

Shiromizu, Tetsuya; Shibata, Masaru Black holes in the brane world: Time symmetric initial data Journal Physical Review Date: Dec 15 2000 Abstract: Abstract available at publisher web site Journal volume: 62 Issue number: 12 Page number: 127502-127502-4 Publisher: American Physical Society

Shmakova, M. Calabi-Yau black holes Journal Physical Review, D (Particles Fields) Date: Jul 31 1997 Abstract: The authors find the entropy of N=2 extreme black holes associated with general Calabi-Yau moduli space and the prepotential $F=d_{ABC}(X^A X^B X^C/X^0)$. The authors show that for arbitrary d_{ABC} and black hole charges p and q the entropy-area formula depends on combinations of these charges and parameters d_{ABC}. These combinations are the solutions of a simple system of algebraic equations. Journal volume: 56 Issue number: 2 Page number: R540-R544 Subjects: Supergravity; Entropy; Charges; Supersymmetry

Shojai, Ali Quantum, Gravity and Geometry Journal International Journal of Modern Physics A Date: May 10 2000 Abstract: Recently,1-3 it is shown that, the quantum effects of matter are well described by the conformal degree of freedom of the space-time metric. On the other hand, it is a well-known fact that according to Einstein's gravity theory, gravity and geometry are interconnected. In the new quantum gravity theory,1-3 matter quantum effects completely determine the conformal degree of freedom of the space-time metric, while the causal structure of the space-time is determined by the gravitational effects of the matter, as well as the quantum effects through back reaction effects. This idea, previously, is realized in the framework of scalar-tensor theories.

Sigurdsson, Steinn Estimating the detectable rate of capture of stellar mass black holes by massive central black holes in normal galaxies Journal Classical and Quantum Gravity Date: Jun 01 1997 Abstract: The capture and subsequent inspiral of stellar mass black holes on eccentric orbits by central massive black holes is one of the more interesting likely sources of gravitational radiation detectable by LISA. The authors estimate the rate of observable events and the associated uncertainties. A moderately favourable mass function could provide many detectable bursts each year, and a detection of at least one burst per year is very likely given our current understanding of the populations in cores of normal spiral galaxies. Journal volume: 14 Issue number: 6 Page number: 1425-1429 Publisher: Institute of Physics

Singh, T. P.; Vaz, Cenalo Quantum radiation from black holes and naked singularities in spherical dust collapse Journal Physical Review Date: Jun 15 2000 Abstract: Abstract available at publisher web site Journal volume: 61 Issue number: 12 Page number: 124005-124005-11 Publisher: American Physical Society

Siopsis, G. Newtonian versus black-hole scattering Journal Physical Review, D (Particles Fields) Date: Feb 28 1999 Abstract: The authors discuss non-relativistic scattering by a Newtonian potential. The authors show that the graybody factors associated with scattering by a black hole exhibit the same functional dependence as scattering amplitudes in the Newtonian limit, which should be the weak-field limit of any quantum theory of gravity. This behavior arises independently of the presence of supersymmetry. Journal volume: 59 Issue number: 4 Page number: 044015 Subjects: Potential Scattering; Quantum Gravity; Conformal Invariance

Siopsis, George Quantization of maximally charged slowly moving black holes Journal Physical Review Date: May 15 2001 Abstract: The authors discuss the quantization of a system of slowly moving extreme Reissner-Nordström black holes. In the near-horizon limit, this system has been shown to possess an SL(2,R) conformal symmetry. However, the Hamiltonian appears to have no well-defined ground state. This problem can be circumvented by a redefinition of the Hamiltonian due to de Alfaro, Fubini, and Furlan (DFF). The authors apply the Faddeev-Popov quantization procedure to show that the Hamiltonian with no ground state corresponds to a gauge in which there is an obstruction at the singularities of moduli space requiring a modification of the quantization rules.

Siopsis, George Quantization of maximally charged slowly moving black holes Journal Physical Review Date: May 15 2001 Abstract: Abstract available at publisher web site Journal volume: 63 Issue number: 10 Page number: 104018-104018-10 Publisher: American Physical Society

Siopsis, George Scalar absorption by spinning D3-branes Journal Physical Review. D, Particles Fields Date: Jan 15 2000 Abstract: The authors discuss absorption of scalars by a distribution of spinning D3-branes. The D3-branes are multi-center solutions of supergravity theory. The authors solve the wave equation in various cases of supergravity backgrounds in which the equation becomes separable. The authors show that the absorption coefficients exhibit a universal behavior as functions of the angular momentum quantum number and the Hawking temperature.

Siopsis, George Scalar absorption by spinning D3-branes Journal Physical Review Date: Jan 15 2000 Abstract: Abstract available at publisher web site Journal volume: 61 Issue number: 2 Page number: 024029-024029-11 American Physical Society

Sivaram, C. Comment on "Thermodynamics of Black Holes: An Analogy with Glasses" Journal Physical Review Letters Date: Apr 03 2000 Abstract: Abstract available at publisher web site Journal volume: 84 Issue number: 14 Page number: 3209-3209 Publisher: American Physical Society

Smalley, Larry L.; Krisch, Jean P. Spinning string fluid dynamics in general relativity Journal Classical and Quantum Gravity Date: Dec 01 1997 Abstract: The general, energy - momentum tensor for a dynamical, spinning string fluid in general relativity is presented using the Ray - Hilbert variational principle. The calculations are given for both the standard and the extended thermodynamics versions in which the latter includes both the spin and string as thermodynamic variables. Both versions yield the same Fermi - Walker transport of the spin with a correction term due to the string. As an unexpected general feature, it is shown that the string and spin are dual 2-forms. Examples of solutions to the field equations are given for the extension of static black holes for non-spinning, string fluids to stationary, slowly rotating black holes for spinning string fluids. These solutions are then compared with solutions obtained from the postulated energy - momentum tensor of Letelier. The general feature of these solutions for positive density is that the spin causes the event horizon for the stationary black hole to contract whether or not the spin is considered as a thermodynamic variable. Journal volume: 14 Issue number: 12 Page number: 3501-3512 Publisher: Institute of Physics

Smoller, J. A.; Wasserman, A. G. Extendability of Solutions of the Einstein-Yang/Mills Equations Journal Communications in Mathematical Physics Date: Jun 01 1998 Abstract: Abstract:The authors prove that any solution to the spherically symmetric SU(2) Einstein-Yang/Millsequations that is defined in the far field and is asymptotically flat, is globally defined. This result applies in particular to the interior of colored black holes. Publisher: Springer-Verlag New York, Inc.

Smoller, J.A; Wasserman, A.G. Reissner–Nordström-like solutions of the SU(2) Einstein–Yang/Mills equations Journal Journal of Mathematical Physics (New York) Date: Dec 31 1997 Abstract: The authors introduce a new class of spherically symmetric solutions of the SU(2) Einstein–Yang/Mills equations. These solutions have a Reissner–Nordström-type essential singularity at the origin, and are well behaved in the far field. These solutions are needed to classify all spherically symmetric solutions which are smooth, asymptotically flat in the far field, and have finite (ADM) mass. Journal volume: 38 Issue number: 12 Page number: 6522-6559 Subject Keyword: Einstein Field Equations/Yang-Mills Theory; Su-2 Groups; Singularity

Smoller, Joel; Temple, Blake On the Oppenheimer-Volkoff Equations in General Relativity Journal Archive for Rational Mechanics and Analysis Date: May 01 1998 Abstract: Abstract.The authors introduce a new formulation of the Oppenheimer-Volkoff (O-V) equations, a system of ordinary differential equations that models the interior of a star in general relativity, and the authors use this to give a completely rigorous mathematical analysis of solutions. In particular, the authors prove that, under mild assumptions on the equation of state, black holes never form in solutions of the O-V equations. As a corollary, this implies that the portion of the empty-space Schwarzschild solution inside the Schwarzschild radius cannot be obtained as a limit of O-V solutions having non-zero density.

Sonoda, H. Understanding the chiral anomaly in coordinate space Journal Physical Review, D (Particles Fields) Date: Apr 30 1997 Abstract: By completing the old discussion of Wilson, the authors express the chiral anomaly in terms of a double integral of a three-point function of chiral currents over an arbitrarily small region in coordinate space. An integrability condition provides an important finite local counterterm to the integral. Journal volume: 55 Issue number: 8 Page number: 5245-5247 Subjects: Numerical Analysis; Gravitation; Einstein Field Equations; Coordinates; Schwinger Functional Equations; Singularity; Quantum Chromodynamics; Su Groups

Sorkin, Evgeny; Piran, Tsvi Formation and evaporation of charged black holes Journal Physical Review Date: Jun 15 2001 Abstract: The authors investigate the dynamical formation and evaporation of a spherically symmetric charged black hole. The authors study the self-consistent one loop order semiclassical back reaction problem. To this end the mass evaporation is modeled by an expectation value of the stress-energy tensor of a neutral massless scalar field, while the charge is not radiated away. The authors observe the formation of an initially nonextremal black hole which tends toward the extremal black hole $M=Q$, emitting a Hawking radiation. Journal volume: 63 Issue number: 12 Page number: 124024-124024-10 Publisher: American Physical Society

Starkman, Glenn D.; Stojkovic, Dejan How frustrated strings would pull the black holes from the centers of galaxies Journal Physical Review Date: Feb 15 2001 Abstract: Abstract available at publisher web site Journal volume: 63 Issue number: 4 Page number: 045008-045008-5 Publisher: American Physical Society

Stone, James M.; Pringle, James E. Magnetohydrodynamical non-radiative accretion flows in two dimensions Journal Monthly Notices of the Royal Astronomical Society Date: Apr 01 2001 Abstract: The authors present the results of axisymmetric, time-dependent magnetohydrodynamic simulations of accretion flows around black holes. The calculations begin from a rotationally supported thick torus which contains a weak poloidal field. Accretion is produced by growth and saturation of the magnetorotational instability (MRI) provided that the wavelength of the fastest growing mode is less than the thickness of the torus. Using a computational grid that spans more than two decades in radius, the authors compare the time-averaged properties of the flow with previous hydrodynamical simulations.

Stryla, Jerzy Isometric embeddings of the reduced horizon of a Kerr-Newman black hole into Euclidean 3-space Journal Classical and Quantum Gravity Date: Aug 01 1998 Abstract: The authors prove some results concerning Killing fields, isometric immersions and black hole horizons, and then combine them to arrive at conclusions concerning embeddings of reduced black hole horizons into Euclidean space. The authors show that not far from the Schwarzschild solution in the class of Kerr-Newman solutions of Einstein equations the reduced horizon of a black hole has a unique isometric embedding in . But certain critical values of parameters describing this class exist, for which it does not take place; when the reduced horizon has negative Gauss curvature areas. The authors compare these parameters with the thermodynamics of black holes. Journal volume: 15 Issue number: 8 Page number: 2303-2310 Publisher: Institute of Physics

Surya, Sumati; Schleich, Kristin; Witt, Donald M. Phase Transitions for Flat Anti–de Sitter Black Holes Journal Physical Review Letters Date: Jun 04 2001 Abstract: The authors reexamine the thermodynamics of anti–de Sitter (adS) black holes with Ricci flat horizons using the adS soliton as the thermal background. The authors find that there is a phase transition which is dependent not only on the temperature but also on the black hole area, which is an independent parameter. As in the spherical adS black hole, this phase transition is related via the adS/conformal-field-theory correspondence to a confinement-deconfinement transition in the large-N gauge theory on the conformal boundary at infinity. Journal volume: 86 Issue number: 23 Page number: 5231-5234 Publisher: The American Physical Society

Suzuki, K. Nonextremal stringy black hole Journal Physical Review, D (Particles Fields) Date: Dec 31 1997 Abstract: The authors construct a four-dimensional BPS saturated heterotic string solution from the Taub-NUT solution. It is a nonextremal black hole solution since its Euler number is nonzero. The authors evaluate its black hole entropy semiclassically. The authors discuss the relation between the black hole entropy and the degeneracy of string states. Journal volume: 56 Issue number: 12 Page number: 7846-7853 Subjects: Black Holes/String Models; Four-Dimensional Calculations; Entropy; Semiclassical Approximation

Szuszkiewicz, Ewa; Miller, John C. Non-linear evolution of thermally unstable slim accretion discs with a diffusive form of viscosity Journal Monthly Notices of the Royal Astronomical Society Date: Nov 01 2001 Abstract: The authors are carrying out a programme of non-linear, time-dependent numerical calculations to study the evolution of the thermal instability driven by radiation pressure in transonic accretion discs around black holes. In our previous studies the authors first investigated the original version of the slim-disc model with low viscosity &formmul; for a stellar-mass (10 M) black hole, comparing the behaviour seen with results from local stability analysis (which were broadly confirmed).

Tamaki, Takashi; Maeda, Kei-ichi Estimating Hawking radiation for exotic black holes

Journal Physical Review Date: Nov 15 2000 Abstract: Abstract available at publisher web site Journal volume: 62 Issue number: 10 Page number: 107503-107503-4 Publisher: American Physical Society

Tamaki, Takashi; Maeda, Kei-ichi Fate of a Reissner-Nordström black hole in the Einstein-Yang-Mills-Higgs system Journal Physical Review Date: Oct 15 2000 Abstract: Abstract available at publisher web site Journal volume: 62 Issue number: 8 Page number: 084041-084041-8 Publisher: American Physical Society

Tamaki, Takashi; Maeda, Kei-ichi; Inada, Masakazu Properties of black hole solutions in the SU(3) Einstein-Yang-Mills-dilaton system Journal Physical Review Date: Apr 15 2001 Abstract: Abstract available at publisher web site Journal volume: 63 Issue number: 8 Page number: 087504-087504-4 Publisher: American Physical Society

Tamaki, Takashi; Torii, Takashi Dyonic BIon black hole in string inspired model Journal Physical Review Date: Jul 15 2001 Abstract: Abstract available at publisher web site The critical BI parameter divides the solutions qualitatively. For there exists a particle like solution for which the dilaton field is finite everywhere, while a no particlelike solution exists and the solution in the rh limit becomes naked. Though there is an extreme solution, the BPS saturated solution does not exist in this case. When the solutions have both magnetic and electric charge, the authors obtain the nontrivial axion field which plays an important role particularly for small black holes. The thermodynamical properties and the configuration of the dilaton field approach those in the magnetically charged case in the zero horizon limit, although the gravitational mass does not. This is related to the nontrivial behavior of the axion field. The authors can prove that there is no inner horizon and that the global structure is the same as the Schwarzschild black hole in any charged case. Journal volume: 64 Issue number: 2 Page number: 024027-024027-11 Publisher: American Physical Society

Tamaki, Takashi; Torii, Takashi Gravitating BIon and BIon black hole with a dilaton Journal Physical Review Date: Sep 15 2000 Abstract: Abstract available at publisher web site Journal volume: 62 Issue number: 6 Page number: 061501-061501-5 Publisher: American Physical Society

Tamaki, Takashi; Yajima, Hiroki Thermodynamic properties of massive dilaton black holes Journal Physical Review Date: Oct 15 2001 Abstract: Abstract available at publisher web site Journal volume: 64 Issue number: 8 Page number: 084002-084002-9 Publisher: American Physical Society

Taylor, Brett E.; Hiscock, William A.; Anderson, Paul R. Semiclassical charged black holes with a quantized massive scalar field Journal Physical Review Date: Apr 15 2000 Abstract: Abstract available at publisher web site Journal volume: 61 Issue number: 8 Page number: 084021-084021-7 Publisher: American Physical Society

Team, K Danzmann.for.the.LISA.Study LISA - an ESA cornerstone mission for a gravitational wave observatory Journal Classical and Quantum Gravity Date: Jun 01 1997 Abstract: The European Space Agency has selected LISA, a gravitational wave observatory, as a cornerstone mission in its future science program Horizons 2000. This observatory will complement the development of ground-based gravitational wave detectors currently under construction. A spaceborne detector will enable the observation of low-frequency gravitational waves in a frequency range from to which is totally inaccessible to ground-based experiments. This frequency range is unique in that it is expected to contain signals from massive black holes, galactive binary stars, as well as the most violent events in the Universe. P LISA will attain this low-frequency sensitivity by employing laser interferometric distance measurements over a very long baseline of . Three of these baselines form an equilateral triangle with spacecraft at each vertex. The cluster of spacecraft is in an Earth-like orbit around the Sun trailing the Earth by . P The spacecraft contain infrared light-emitting Nd:YAG lasers and freely floating test masses made from a special platinum - gold alloy with vanishing magnetic susceptibility. The spacecraft are being kept centred on their test masses by using drag-free technology and field-emission electric propulsion, thus letting the test masses follow purely inertial orbits. Journal volume: 14 Issue number: 6 Page number: 1399-1404 Publisher: Institute of Physics

Tichy, Wolfgang; Flanagan, Éanna É Angular momentum ambiguities in asymptotically flat spacetimes which are perturbations of stationary spacetimes Journal Classical and Quantum Gravity Date: Oct 07 2001 Abstract: P The authors use simple scaling and symmetry arguments to deduce the following properties of spacetimes consisting of point test particles orbiting Kerr black holes. (a) The angular momentum at spatial and null infinity, and the angular momentum flux at null infinity, are well defined in the sense that the ambiguities in these quantities are smaller than the quantities themselves by one or more factors of the dimensionless mass ratio (mass of test particle)/(mass of black hole). (b) If the total angular momentum of the spacetime is taken to point in the z direction, then the x and y components of angular momentum flux vanish, and consequently the vectorial angular momentum flux does not yield enough information to infer the radiation-reaction driven evolution of the Carter constant in the adiabatic regime.

Tim de Zeeuw, P.; Bureau, M.; Emsellem, Eric; Bacon, R.; Marcella Carollo, C.; Copin, Y.; Davies, Roger L.; Kuntschner, Harald; Miller, Bryan W.; Monnet, G.; Peletier, Reynier F.; Verolme, E.K. The SAURON project - II. Sample and early results Journal Monthly Notices of the Royal Astronomical Society Date: Jan 01 2002 Abstract: Early results are reported from the SAURON survey of the kinematics and stellar populations of a representative sample of nearby E, S0 and Sa galaxies. The survey is aimed at determining the intrinsic shape of the galaxies, their orbital structure, the mass-to-light ratio as a function of radius, the age and metallicity of the stellar populations, and the frequency of kinematically decoupled cores and nuclear black holes. The construction of the representative sample is described, and its properties are illustrated.

Tomimatsu, Akira; Koyama, Hiroko Vacuum polarization of scalar fields near Reissner-Nordström black holes and the resonance behavior in field-mass dependence Journal Physical Review Date: Jun 15 2000 Abstract: Abstract available at publisher web site Journal volume: 61 Issue number: 12 Page number: 124010-124010-10 Publisher: American Physical Society

Umeda, Hideyuki; Nomoto, Ken'ichi; Tsuruta, Sachiko; Mineshige, Shin Impacts of the Detection of Cassiopeia A Point Source Journal The Astrophysical Journal Date: Apr 01 2000 Abstract: Very recently the Chandra first light observation discovered a point-like source in the Cassiopeia A supernova remnant. This detection was subsequently confirmed by the analyses of the archival data from both ROSAT and Einstein observations. Here the authors compare the results from these observations with the scenarios involving both black holes (BHs) and neutron stars (NSs). If this point source is a BH, the authors offer as a promising model a disk-corona type model with a low accretion rate in which a soft photon source at ~0.1 keV is Comptonized by higher energy electrons in the corona.

Valtancoli, P. Integrability of the N-Body Problem in (2+1)-AdS Gravity Journal International Journal of Modern Physics A Date: Oct 30 2000 Abstract: The authors derive a first order formalism for solving the scattering of point sources in (2+1) gravity with negative cosmological constant. We show that their physical motion can be mapped, with a polydromic coordinate transformation, to a trivial motion, in such a way that the point sources move as timelike geodesics (in the case of particles) or as spacelike geodesics (in the case of BTZ black holes) of a three-dimensional hyper surface immersed in a four-dimensional Minkowskian space-time, and that the two-body dynamics is solved by two invariant masses, whose difference is simply related to the total angular momentum of the system. Journal volume: 15 Issue number: 27 Rights: World Scientific Publishing Company Publisher: World Scientific Publishing Company

van der Bij, J. J.; Radu, Eugen Gravitating sphalerons and sphaleron black holes in asymptotically anti-de Sitter spacetime Journal Physical Review Date: Sep 15 2001 Abstract: Abstract available at publisher web site Journal volume: 64 Issue number: 6 Page number: 064020-064020-11 American Physical Society

van Nieuwenhuizen, Peter; Nojiri, Shin'ichi; Odintsov, Sergei D. Conformal anomaly for 2D and 4D dilaton coupled spinors Journal Physical Review Date: Oct 15 1999 Abstract: Abstract available at publisher web site Journal volume: 60 Issue number: 8 Page number: 084014-084014-7 Publisher: American Physical Society

van Putten, Maurice H. P. M. Electron-Positron Outflow from Black Holes Journal Physical Review Letters Date: Apr 24 2000 Abstract: Abstract available at publisher web site Journal volume: 84 Issue number: 17 Page number: 3752-3755 Publisher: American Physical Society

van Putten, Maurice H. P. M.; Levinson, Amir Detecting Energy Emissions from a Rotating Black Hole Journal Title: Science Magazine Date: Mar 08 2002 Abstract: Abstract available at publisher web site Journal volume: 295 Issue number: 5561 Page number: 1874-1877 DOE sponsorship Publisher: American Association for the Advancement of Science

van Putten, Maurice H. P. M.; Sarkar, Abhinanda An f˙(f)-frequency dynamics algorithm for gravitational waves Journal Physical Review Date: Aug 15 2000 Abstract: Abstract available at publisher web site Journal volume: 62 Issue number: 4 Page number: 041502-041502-5 Publisher: American Physical Society

Varadarajan, Madhavan Kruskal coordinates as canonical variables for Schwarzschild black holes Journal Physical Review Date: Apr 15 2001 Abstract: Abstract available at publisher web site Journal volume: 63 Issue number: 8 Page number: 084007-084007-8 Publisher: American Physical Society

Vaz, Cenalo; Witten, Louis Quantum black holes from quantum collapse Journal Physical Review Date: Oct 15 2001 Abstract: Abstract available at publisher web site Journal volume: 64 Issue number: 8 Page number: 084005-084005-6 Publisher: American Physical Society

Vaz, Cenalo; Witten, Louis Quantum states and the statistical entropy of the charged black hole Journal Physical Review Date: Jan 15 2001 Abstract: The authors quantize the Reissner-Nordström black hole using an adaptation of Kuchar's canonical decomposition of the Kruskal extension of the Schwarzschild black hole. The Wheeler-DeWitt equation turns into a functional Schrödinger equation in Gaussian time by coupling the gravitational field to a reference fluid or dust. The physical phase space of the theory is spanned by the mass M, the charge Q, the physical radius R, the dust proper time {tau}, and their canonical momenta. The exact solutions of the functional Schrödinger equation imply that the difference in the areas of the outer and inner horizons is quantized in integer units.

Vaz, Cenalo; Witten, Louis Quantum states and the statistical entropy of the charged black hole Journal Physical Review Date: Jan 15 2001 Abstract: Abstract available at publisher web site Journal volume: 63 Issue number: 2 Page number: 024008-024008-11 Publisher: American Physical Society

Vaz, Cenalo; Witten, Louis; Singh, T. P. Toward a midi superspace quantization of LeMaitre-Tolman-Bondi collapse models Journal Physical Review Date: May 15 2001 Abstract: LeMaitre-Tolman-Bondi models of spherical dust collapse have been used and continue to be used extensively to study various stellar collapse scenarios. It is by now well known that these models lead to the formation of black holes and naked singularities from regular initial data. The final outcome of the collapse, particularly in the event of naked singularity formation, depends very heavily on quantum effects during the final stages. These quantum effects cannot generally be treated semiclassically as quantum fluctuations of the gravitational field are expected to dominate before the final state is reached.

Vendrell, F. Quantum mechanical path integrals and thermal radiation in static curved spacetimes Journal Physical Review Date: Feb 15 2000 Abstract: Abstract available at publisher web site Journal volume: 61 Issue number: 4 Page number: 044019-044019-17 Publisher: American Physical Society

Venkatesan, A.; Olinto, A.V.; Truran, J.W. Neutron Stars and Black Holes as MACHO{bold s} Journal Astrophysical Journal Date: May 01 1999 Abstract: The authors consider the contribution of neutron stars and black holes to the dynamical mass of galactic halos. In particular, the authors show that if these compact objects were produced by an early generation of stars with initial metallicity {approx_lt}10{sup {minus}4} Z {sub {circle_dot}}, they can contribute at most 30%–40% of the Galactic halo mass without creating supersolar levels of enrichment. The authors show that the case for halo neutron stars and black holes cannot be rejected on metal overproduction arguments alone because of the critical factor of the choice of progenitor metallicity in determining the yields. The authors show that this scenario satisfies observational constraints, similar to but no more severe than those faced by halo white dwarfs. The authors also discuss the recent results on halo microlensing, the presence of enriched hot gas in clusters and groups of galaxies, and other observations. If there are halo neutron stars and black holes, they will be detected in the future as longer timescale events by microlensing experiments. Journal volume: 516 Issue number: 2 Page number: pp.863-871

Verbunt, Frank; Nelemans, Gijs Binaries for LISA Journal Classical and Quantum Gravity Date: Oct 07 2001 Abstract: P Binaries sufficiently compact to be observed with LISA consist of small stars, with both chosen from the following list: low-mass main sequence stars, white dwarfs, neutron stars, black holes or helium-burning cores of high-mass stars. The numbers of various types of binaries can be derived from observations, provided that observational selection effects are taken into account. They can also be derived from theoretical population synthesis, but the uncertainties are such that the resulting numbers must be gauged with observed samples. Journal volume: 18 Issue number: 19 Page number: 4005-4011 Publisher: Institute of Physics

Villiers, Jean-Pierre De; Frolov, Valeri Gravitational scattering of cosmic strings by non-rotating black holes Journal Classical and Quantum Gravity Date: Jul 01 1999 Abstract: This paper discusses the gravitational scattering of a straight, infinitely long test cosmic string by a black hole. The authors present numerical results that probe the two-dimensional parameter space of impact parameter and initial velocity and compare them to approximate perturbative solutions derived previously. The authors analyse string scattering and coil formation in the ultra-relativistic regime and compare these results with analytical results for string scattering by a gravitational shock wave. Special attention is paid to regimes where the string approaches the black hole at near-critical impact parameters

Visser, M. Hawking Radiation without Black Hole Entropy Journal Physical Review Letters Date: Apr 30 1998 Abstract: Hawking radiation is a purely kinematic effect that is generic to Lorentzian geometries containing event horizons; it is independent of dynamics. On the other hand, the classical laws of black hole mechanics and the semiclassical laws of black hole thermodynamics are both inextricably linked with dynamics: Black hole entropy is proportional to area (plus corrections) if and only if the dynamics is Einstein-Hilbert (plus corrections). Hawking radiation can occur in physical situations in which the laws of black hole mechanics do not apply, and in physical situations in which the notion of black hole entropy does not even make any sense. Journal volume: 80 Issue number: 16 Page number: 3436-3439 Subjects: Einstein Field Equations; Geometry; General Relativity Theory; Quantum Gravity; Perturbation Theory

Visser, Matt Acoustic black holes: horizons, ergospheres and Hawking radiation Journal Classical and Quantum Gravity Date: Jun 01 1998 Abstract: It is a deceptively simple question to ask how acoustic disturbances propagate in a non-homogeneous flowing fluid. Subject to suitable restrictions, this question can be answered by invoking the language of Lorentzian differential geometry. This paper begins with a pedagogical derivation of the following result: if the fluid is barotropic and inviscid, and the flow is irrotational (though possibly time dependent), then the equation of motion for the velocity potential describing a sound wave is identical to that for a minimally coupled massless scalar field propagating in a $(3 + 1)$-dimensional Lorentzian geometry.

Wadia, Spenta R. Black Holes, Information Puzzle and String Theory Journal International Journal of Modern Physics B Date: Aug 10 2000 Abstract: The authors briefly review developments in string theory which explain the information puzzle that arises in the applications of quantum mechanics to black holes. Journal volume: 14 Issue number: 19_20 Publisher: World Scientific Publishing Company

Wagner, S.; Adams, H.-P.; Sobel, D.F.; Slivka, L.S.; Sipe, J.C.; Romine, J.S.; Koziol, J.A. New Hypointense Lesions on MRI in Relapsing-Remitting Multiple Sclerosis Patients Journal European Neurology Date: May 01 2000 Abstract: Background: Preliminary observational studies with multiple sclerosis (MS) patients have reported strong correlations between an increase in hypointense lesion load (black holes) on T1-weighted spin echo images, and an increase in disability. Objective: The authors assessed the relationship of hypointense lesions to the clinical course of disease among 50 relapsing-remitting MS patients in the controlled setting of a randomized clinical trial.

Wang, Bin; Abdalla, Elcio Entropy bound for a rotating system from anti-de Sitter black holes Journal Physical Review Date: Aug 15 2000 Abstract: Abstract available at publisher web site Journal volume: 62 Issue number: 4 Page number: 044030-044030-5 American Physical Society

Wang, Bin; Abdalla, Elcio; Su, Ru-Keng Friedmann Equation and Cardy Formula Correspondence in Brane Universes Journal Modern Physics Letters A Date: Jan 01 2002 Abstract: The authors study the brane with arbitrary tension on the edge of various black holes with AdS asymptotics. The authors investigate Friedmann equations governing the motion of the brane universes and match the Friedmann equation to Cardy entropy formula. Journal volume: 17 Issue number: 01 Publisher: World Scientific Publishing Company

Wang, Bin; Abdalla, Elcio; Su, Ru-Keng Geometry and topology of two kinds of extreme Reissner-Nordström anti-de Sitter black holes Journal Physical Review Date: Aug 15 2000 Abstract: Abstract available at publisher web site Journal volume: 62 Issue number: 4 Page number: 047501-047501-4 American Physical Society

Wang, Bin; Su, Ru-Keng; Abdalla, Elcio Extreme Black Hole Entropy Obtained in an Operational Approach Journal International Journal of Modern Physics A Date: Mar 01 2001 Abstract: The entropy of anti-de Sitter Reissner-Nordström black hole is found to be stored in the material which gathers to form it and equals to A/4 regardless of material states. Extending the study to two kinds of extreme black holes, the authors find different entropy results for the first kind of extreme black hole due to different material states. However for the second kind of extreme black hole the results of entropy are uniform independently of the material states. Relations between these results and the stability of two kinds of extreme black holes have been addressed. Journal volume: 16 Issue number: 08 Publisher: World Scientific Publishing Company

Wardzinski, Grzegorz; Zdziarski, Andrzej A. Effects of non-thermal tails in Maxwellian electron distributions on synchrotron and Compton processes Journal Monthly Notices of the Royal Astronomical Society Date: Aug 01 2001 Abstract: The authors investigate how the presence of a non-thermal tail beyond a Maxwellian electron distribution affects the synchrotron process as well as Comptonization in plasmas with parameters typical for accretion flows on to black holes. The authors find that the presence of the tail can significantly increase the net (after accounting for self-absorption) cyclo-synchrotron emission of the plasma, which then provides seed photons for Compton upscattering. Thus, the luminosity in the thermally Comptonized spectrum is enhanced as well.

Weinberg, Martin D. Noise-driven evolution in stellar systems - II. A universal halo profile Journal Monthly Notices of the Royal Astronomical Society Date: Nov 01 2001 Abstract: Disc instabilities such as arm and bar formation, minor mergers and tidal encounters drive a galaxy from equilibrium. Using the theory that describes the evolution of a galaxy halo as a result of stochastic fluctuations developed in the companion paper to this one, the authors show that this sort of noise evolves a halo toward a standard profile, independent of its initial profile and concentration. This process can substantially redistribute the mass in dark-matter haloes in the 10 Gyr since formation. Three different noise processes are studied: (i) a bombardment by blobs of mass that are small compared to the halo mass ('shrapnel'); (ii) orbital evolution of substructure by dynamical friction ('satellites') and (iii) noise caused by the orbit of blobs in the halo ('black holes').

Wessels, Ewald Exact analytic characteristic initial data for axi symmetric, non-rotating, vacuum spacetimes, with an application to the binary black hole problem Journal Classical and Quantum Gravity Date: Aug 01 1998 Abstract: Bondi's approach to the construction of a coordinate system is used with a different choice of gauge, in accordance with which the radial coordinate r is an affine parameter, to cast the metric tensor into a form suitable for use with the Newman-Penrose null tetrad formalism. The choice of tetrad has the result that the equations and all the functions that appear in them are real-valued. A group classification of the Sachs equations in this gauge leads to a unique expression for the first of the five independent elements of the Weyl spinor, and to the corresponding exact solutions for two of the metric functions on an initial null hypersurface. A proof is presented that the result for constitutes the appropriate characteristic initial value function for all physically realistic axisymmetric, non-rotating vacuum spacetimes.

Wilkinson, P. N.; Henstock, D. R.; Browne, I. W. A.; Polatidis, A. G.; Augusto, P.; Readhead, A. C. S.; Pearson, T. J.; Xu, W.; Taylor, G. B.; Vermeulen, R. C. Limits on the Cosmological Abundance of Supermassive Compact Objects from a Search for Multiple Imaging in Compact Radio Sources Journal Physical Review Letters Date: Jan 22 2001 Abstract: Abstract available at publisher web site Journal volume: 86 Issue number: 4 Page number: 584-587 Publisher: American Physical Society

Wilman, R. J.; Fabian, A. C.; Nulsen, P. E. J. A physical model for the hard X-ray background Journal Monthly Notices of the Royal Astronomical Society Date: Dec 01 2000 Abstract: The authors combine a semi-analytic galaxy formation model with a prescription for the obscured growth of massive black holes, to reproduce the hard X-ray background (XRB), the local 2-10 keV active galactic nuclei (AGN) luminosity function and the source counts, including recent Chandra results. The model also complies with constraints on the AGN contribution to the far-infrared and submillimetre backgrounds.

Wilms, J.; Nowak, M. A.; Pottschmidt, K.; Heindl, W. A.; Dove, J. B.; Begelman, M. C. Discovery of recurring soft-to-hard state transitions in LMC X-3 Journal Monthly Notices of the Royal Astronomical Society Date: Jan 01 2001 Abstract: The authors present the analysis of the approximately three-year long Rossi X-ray Timing Explored (RXTE) campaign for monitoring the canonical soft state black-hole candidates LMC X-1 and LMC X-3. In agreement with previous observations, the authors find that the spectra of both sources can be well-described by the sum of a multi-temperature disc blackbody and a power law. In contrast to LMC X-1, which does not exhibit any periodic spectral changes, the authors find that LMC X-3 exhibits strong spectral variability on time-scales of days to weeks.

Winstanley, E. Existence of stable hairy black holes in su(2) Einstein-Yang-Mills theory with a negative cosmological constant Journal Classical and Quantum Gravity Date: Jun 01 1999 Abstract: The authors consider black holes in Einstein-Yang-Mills theory with a negative cosmological constant. The solutions obtained are somewhat different from those for which the cosmological constant is either positive or zero. Firstly, regular black hole solutions exist for continuous intervals of the parameter space, rather than discrete points. Secondly, there are non-trivial solutions in which the gauge field has no nodes. The authors show that these solutions are linearly stable. Journal volume: 16 Issue number: 6 Page number: 1963-1978 Publisher: Institute of Physics

Winstanley, Elizabeth Classical super-radiance in Kerr-Newman-anti-de Sitter black holes Journal Physical Review Date: Nov 15 2001 Abstract: Abstract available at publisher web site Journal volume: 64 Issue number: 10 Page number: 104010-104010-10 American Physical Society

Winstanley, Elizabeth Renormalized black hole entropy in anti-de Sitter space via the "brick wall" method Journal Physical Review Date: Apr 15 2001 Abstract: Abstract available at publisher web site Journal volume: 63 Issue number: 8 Page number: 084013-084013-13 Publisher: American Physical Society

Wirschins, Marion; Sood, Abha; Kunz, Jutta Non-Abelian Einstein-Born-Infeld black holes Journal Physical Review Date: Apr 15 2001 Abstract: Abstract available at publisher web site Journal volume: 63 Issue number: 8 Page number: 084002-084002-5 Publisher: American Physical Society

Wiseman, Alan G. Self-force on a static scalar test charge outside a Schwarzschild black hole Journal Physical Review Date: Apr 15 2000 Abstract: Abstract available at publisher web site Journal volume: 61 Issue number: 8 Page number: 084014-084014-14 Publisher: American Physical Society

Witten, Edward Anti-de Sitter Space, Thermal Phase Transition and Confinement in Gauge Theories Journal International Journal of Modern Physics A Date: Jun 01 2001 Abstract: The correspondence between supergravity (and string theory) on AdS space and boundary conformal field theory relates the thermodynamics of N=4 super-Yang-Mills theory in four dimensions to the thermodynamics of Schwarzschild black holes in anti-de Sitter space. In this description, quantum phenomena such as the spontaneous breaking of the center of the gauge group, magnetic confinement and the mass gap are coded in classical geometry. The correspondence makes it manifest that the entropy of a very large AdS Schwarzschild black hole must scale "holographically" with the volume of its horizon. By similar methods, one can also make a speculative proposal for the description of large N gauge theories in four dimensions without supersymmetry. Journal volume: 16 Issue number: 16 Rights: World Scientific Publishing Company Publisher: World Scientific Publishing Company

Wrobel, J. M.; Herrnstein, J. R. Accretion Rates onto Massive Black Holes in Four Quiescent Elliptical Galaxies Journal The Astrophysical Journal Date: Mar 01 2000 Abstract: Four quiescent elliptical galaxies were imaged with the NRAO VLA at 8.5 GHz. Within the context of canonical advection-dominated accretion flows (ADAFs), these VLA images plus published black hole masses constrain the accretion rates to be less than 1.6×10^{-4}, less than 3.6×10^{-4}, 7.8×10^{-4}, and 7.4×10^{-4} of the Eddington rates. These ADAF accretion rates derived at 8.5 GHz have important implications for the levels of soft and hard X-rays expected from these quiescent galaxies. Journal volume: 533 Issue number: 2 Page number: L111-L114 University of Chicago Press

Wu, X. N. Multicritical phenomena of Reissner-Nordström anti-de Sitter black holes Journal Physical Review Date: Dec 15 2000 Abstract: Abstract available at publisher web site Journal volume: 62 Issue number: 12 Page number: 124023-124023-11 American Physical Society

Xu, Dian-Yan Timelike limit surfaces, apparent horizons and event horizons of radiating Kerr-Newman-de Sitter black holes, inflation and cosmic censorship Journal Classical and Quantum Gravity Date: Feb 01 1999 Abstract: Coupling between angular momentum and the cosmological constant does not produce distortion or deformation of apparent horizons. Only to a first-order perturbation or approximation can it affect the shape of event horizons. Journal volume: 16 Issue number: 2 Page number: 343-350 Institute of Physics

Yamaguchi, Masahide Density fluctuations and primordial black hole formation in natural double inflation in supergravity Journal Physical Review Date: Sep 15 2001 Abstract: Abstract available at publisher web site Journal volume: 64 Issue number: 6 Page number: 063503-063503-10 Publisher: American Physical Society

Yamazaki, Ryo; Ida, Daisuke Black holes in three-dimensional Einstein-Born-Infeld-dilaton theory Journal Physical Review Date: Jul 15 2001 Abstract: Abstract available at publisher web site Journal volume: 64 Issue number: 2 Page number: 024009-024009-6 Publisher: American Physical Society

Yazadjiev, Stoytcho S. Distorted charged dilaton black holes Journal Classical and Quantum Gravity Date: Jun 07 2001 Abstract: P The authors construct exact static, axisymmetric solutions of Einstein-Maxwell-dilaton gravity presenting distorted charged dilaton black holes. The thermodynamics of such distorted black holes is also discussed. Journal volume: 18 Issue number: 11 Page number: 2105-2116 Publisher: Institute of Physics

Yazadjiev, Stoytcho S.; Fiziev, Plamen P.; Boyadjiev, Todor L.; Todorov, Michail D. Electrically Charged Einstein-Born Infeld Black Holes with Massive Dilaton Journal Modern Physics Letters A Date: Oct 01 2001 Abstract: The authors numerically construct static and spherically symmetric electrically charged black hole solutions in Einstein-Born-Infeld gravity with massive dilaton. The numerical solutions show that the dilaton potential allows many more black hole causal structures than the massless dilaton. The authors find that depending on the black hole mass and charge and the dilaton mass, the black holes can have either one, two, or three horizons. As an interesting peculiarity the authors note that there are extremal black holes with an inner horizon and with triply degenerated horizon. Journal volume: 16 Issue number: 33 Publisher: World Scientific Publishing Company

Yo, Hwei-Jang; Baumgarte, Thomas W.; Shapiro, Stuart L. Numerical test bed for singularity excision in moving black hole spacetimes Journal Physical Review Date: Dec 15 2001 Abstract: Abstract available at publisher web site Journal volume: 64 Issue number: 12 Page number: 124011-124011-12 Publisher: American Physical Society

Youm, Donam Black hole thermodynamics and two-dimensional dilaton gravity theory Journal Physical Review Date: Feb 15 2000 Abstract: Abstract available at publisher web site Journal volume: 61 Issue number: 4 Page number: 044013-044013-8 American Physical Society

Youm, Donam The Cardy-Verlinde Formula and Asymptotically Flat Charged Black Holes Journal Modern Physics Letters A Date: Jun 01 2001 Abstract: The authors show that the modified Cardy-Verlinde formula without the Casimir effect term is satisfied by asymptotically flat charged black holes in arbitrary dimensions. Thermodynamic quantities of the charged black holes are shown to satisfy the energy-temperature relation of at two-dimensional CFT, which supports the claim in our previous work(Phys. Rev.D61, 044013, hep-th/9910244) that thermodynamics of charged black holes in higher dimensions can be effectively described by two-dimensional theories. The authors also check the Cardy formula for the two-dimensional black hole compactified from adilatonic charged black hole in higher dimensions. Journal volume: 16 Issue number: 19 Rights: World Scientific Publishing Company Publisher: World Scientific Publishing Company

Yu, Qingjuan Evolution of massive binary black holes Journal Monthly Notices of the Royal Astronomical Society Date: Jan 31 2002 Abstract: Abstract available at Blackwell Science. Journal volume: 331 Issue number: 4 Page number: 935-958 DOE sponsorship Subjects: black hole physics; galaxies: evolution; galaxies: interactions; galaxies: kinematics and dynamics; galaxies: nuclei Publisher: Blackwell Science Ltd

Yu, Qingjuan; Lu, Youjun Iron K line profiles driven by non-axisymmetric illumination Journal Monthly Notices of the Royal Astronomical Society Date: Jan 01 2000 Abstract: Previous calculations of Fe K line profiles are based on axisymmetric emissivity laws. In this paper, the authors show line profiles driven by non-axial symmetric illumination which results from an off-axis X-ray point source. The authors find that source location and motion have significant effects on the red wing and blue horn of the line profiles. The disc region under the source will receive more flux, which is the most important factor to affect the line profiles. The authors suggest that at least part of the variation in Fe K line profiles is caused by the motion of X-ray sources. Future observations of Fe K line profiles will provide more information about the distribution and motion of the X-ray sources around black holes, and hence the underlying physics. Journal volume: 311 Issue number: 1 Page number: 161-168 DOE sponsorship Publisher: Blackwell Science

Yuan, Feng Luminous hot accretion discs Journal Monthly Notices of the Royal Astronomical Society Date: Jun 01 2001 Abstract: The authors find a new two-temperature hot branch of equilibrium solutions for stationary accretion discs around black holes. In units of Eddington accretion rate defined as 10L Edd c 2, the accretion rates to which these solutions correspond are within the range &mdot; 1≲&mdot;≲1, where &mdot; 1 is the critical rate of advection-dominated accretion flow (ADAF). In these solutions, the energy loss rate of the ions by Coulomb energy transfer between the ions and electrons is larger than the viscously heating rate and it is the advective heating together with the viscous dissipation that balances the Coulomb cooling of ions.

Yusef-Zadeh, F.; Melia, F.; Wardle, M. The Galactic Center: An Interacting System of Unusual Sources Journal Title: Science Magazine Date: Jan 07 2000 Abstract: Abstract available at publisher web site Journal volume: 287 Issue number: 5450 Page number: 85-91 American Association for the Advancement of Science

Zampieri, Luca; Turolla, Roberto; Szuszkiewicz, Ewa Spectral variability in transonic discs around black holes Journal Monthly Notices of the Royal Astronomical Society Date: Aug 01 2001 Abstract: Transonic discs with accretion rates relevant to intrinsically bright Galactic X-ray sources (L 1038-1039 erg s-1) exhibit a time-dependent cyclic behaviour due to the onset of a thermal instability driven by radiation pressure. In this paper the authors calculate radiation spectra emitted from thermally unstable discs to provide detailed theoretical predictions for observationally relevant quantities. The emergent spectrum has been obtained by solving self-consistently the vertical structure and radiative transfer in the disc atmosphere.

Zaslavskii, O.B. Horizon/matter systems near the extreme state Journal Classical and Quantum Gravity Date: Oct 01 1998 Abstract: It is shown that in the extreme limit with zero surface gravity but non-zero local temperature the limiting metric of a generic static black hole is determined by a metric induced on a horizon and one function of two coordinates, the stress-energy tensor of a source picking up its values from a horizon. The limiting procedure is extended to rotating black holes. If the extreme limit is due to merging a black hole horizon and a cosmological one, both horizons are always in thermal equilibrium in this limit. This is proved for a generic case of static or axially symmetrical rotating spacetimes. Journal volume: 15 Issue number: 10 Page number: 3251-3257 Publisher: Institute of Physics

Zaslavskii, O.B. Non-extreme black holes near the extreme state and acceleration horizons: thermodynamics and quantum-corrected geometry Journal Classical and Quantum Gravity Date: Jan 21 2000 Abstract: The authors consider the class of metrics that can be obtained from those of non-extreme black holes by limiting transitions to the extreme state such that the near-horizon geometry expands into a whole manifold. These metrics include, in particular, the Rindler and Bertotti-Robinson spacetimes. The general formula for the entropy of massless radiation valid either for black hole or for acceleration horizons is derived.

Zdotycki, Piotr T.; Done, Chris; Smith, David A. On the complex disc-corona interactions in the soft spectral states of soft X-ray transients Journal Monthly Notices of the Royal Astronomical Society Date: Oct 01 2001 Abstract: Accreting black holes show a complex and diverse behaviour in their soft spectral states. Although these spectra are dominated by a soft, thermal component which almost certainly arises from an accretion disc, there is also a hard X-ray tail indicating that some fraction of the accretion power is instead dissipated in hot, optically thin coronal material. During such states, best observed in the early outburst of soft X-ray transients, the ratio of power dissipated in the hot corona to that in the disc can vary from ~ 0 (pure disc accretion) to ~ 1 (equal power in each). Here the authors present results of spectral analyses of a number of sources, demonstrating the presence of complex features in their energy spectra.

Zeff, B W; Kleber, B; Fineberg, J; Lathrop, D P
Singularity dynamics in curvature collapse and jet eruption on a fluid surface Journal Nature Date: Jan 27 2000 Abstract: Finite-time singularities—local divergences in the amplitude or gradient of a physical observable at a particular time—occur in a diverse range of physical systems. Examples include singularities capable of damaging optical fibres and lasers in nonlinear optical systems, and gravitational singularities associated with black holes. In fluid systems, the formation of finite-time singularities cause spray and air-bubble entrainment, processes which influence air-sea interaction on a global scale. Singularities driven by surface tension have been studied in the break-up of pendant drops and liquid sheets. Here the authors report a theoretical and experimental study of the generation of a singularity by inertial focusing, in which no break-up of the fluid surface occurs.

Zhu, Jiong-Ming; Wang, Bin; Abdalla, Elcio
Object picture of quasinormal ringing on the background of small Schwarzschild anti–de Sitter black holes Journal Physical Review Date: Jun 15 2001 Abstract: The authors investigate the evolution of a scalar field propagating in small Schwarzschild anti–de Sitter black holes. The imaginary part of the quasinormal frequency decreases with the horizon size and the real part of the quasinormal frequency keeps nearly as a constant. The object picture clarifies the question about the quasinormal modes for small anti–de Sitter black holes. The dependence of the quasinormal modes on spacetime dimensions and the multipole index for small anti–de Sitter black holes is also illustrated. Journal volume: 63 Issue number: 12 Page number: 124004-124004-4. The American Physical Society

AUTHOR INDEX

A

Abdalla, Elcio, 13, 107, 108
Abel, Tom, 59
Abramowicz, Marek A., 13
Achucarro, A., 13
Adams, Fred C., 13, 14
Adams, H.-P S., 107
Agnese, A. G., 14
Akhmedov, E. T., 14
Alcubierre, Miguel, 14
Alexeyev, S. O., 14
Alexeyev, S., 15
Al-Khalili, Jim, 1
Almaini, O., 65
Alvi, Kashif, 15
Åminneborg, Stefan, 15
Anderson, Paul R., 15
Andersson, Nils, 15
Anglin, J. R., 54
Anninos, Peter, 15
Araújo, M.E., 15
Aros, Rodrigo, 16
Arras, Phil, 86
Asensi, Manuel, 5
Ashtekar, Abhay, 16
Ashworth, M. C., 16
Augusto, Pedro, 16
Awad, Adel M., 17
Ayón-Beato, Eloy, 17
Azreg-Aïnou, Mustapha, 18

B

Baaquie, Belal E., 18
Bachas, C., 9
Baker, J., 18
Baker, John, 18

Balasubramanian, V., 18, 19
Balbinot, R., 19
Ballantyne, D. R., 19
Bañados, Máximo, 20
Banks, T., 20
Barack, Leor, 20
Bardeen, James, 21
Bars, I., 21
Barvinsky, Andrei, 21
Bassett, Bruce A., 21
Basu, D., 21
Baumgarte, Thomas W., 22, 90, 98, 111
Becker, K., 22
Beetle, Christopher, 16
Behrndt, K., 22
Bekenstein, Jacob D., 23
Belgiorno, F., 23
Beloborodov, Andrei M., 23, 81
Benger, Werner, 14
Bengtsson, Ingemar, 15
Berezin, Victor, 23
Bergshoeff, Eric, 22
Bernido, C.C., 23, 24
Bertolini, Matteo, 24
Bethe, H.A., 24
Bhattacharyya, S., 25
Bicak, J., 25
Billó, Marco, 25
Birmingham, Danny, 25, 26
Bizon´, Piotr, 26
Bjoraker, Jeff, 26
Bojowald, M., 26
Boldt, Elihu, 26
Bonnor, W.B., 26
Booth, I. S., 27
Borde, A., 27
Borozdin, Konstantin N., 94
Bose, Sukanta, 28
Bousso, Raphael, 28

Boyadjiev, Todor L., 110
Brandenberger, R.H, 79
Brandt, C. F. C., 28
Brandt, Steven, 18, 28
Brandt, W. N., 57
Brax, Philippe, 29
Breckenridge, J.C, 29
Brevik, I., 29
Brihaye, Yves, 29
Brill, Dieter, 15
Bringmann, Torsten, 29
Bromley, B.C., 29
Bronnikov, K. A., 30
Brown, J.D., 30
Brügmann, Bernd, 14
Bullock, J.S., 30
Buonanno, Alessandra, 30
Burko, Lior M., 31
Burnham, Robert, 9

C

Cacciatori, Sergio, 25
Cadeau, C., 31
Cadoni, M., 31, 32
Cai, Mingliang, 32
Cai, Rong-Gen, 33
Caldarelli, Marco M., 33
Campanelli, Manuela, 18
Cardoso, Gabriel Lopes, 34
Cardoso, Vitor, 34
Carlip, S., 34
Carlson, Eric D., 71
Carta, P., 311, 32
Casadio, Roberto, 35
Castiñeiras, J., 35
Cataldo, Mauricio, 35
Cattaneo, Andrea, 36
Celotti, A., 36
Chakrabarti, Sandip K., 6, 36, 37
Chamblin, Andrew, 36, 37
Chamseddine, A., 37
Chartas, George, 96
Chattopadhyay, Indranil, 37
Cherepashchuk, A.M., 38
Chimento, Luis P., 38
Choptuik, Matthew W., 38
Chou, W, 38
Christensen, M., 59
Christodoulakis, T., 38
Chrusciel, Piotr T., 38
Churazov, E., 38
Churches, D., 41

Cirac, J. I., 54
Clément, Gérard, 39
Cognola, Guido, 33
Cohen, Andrew G., 39
Coleman Miller, M., 39
Colgate, S.A., 52
Collins, G.P., 39, 40
Constable, Neil R., 40
Corbin, Michael R., 40
Corichi, Alejandro, 16, 40, 41
Costa, M.S., 41
Coule, D.H., 41
Crisóstomo, Juan, 41
Crispino, Luís C. B., 41
Croce, R. P., 41
Cruz, J. ., 41
Csáki, Csaba, 36
Cui, Wei, 41
Custódio, P. S., 42
Cvetic, Mirjam, 22, 42
Czerny, B., 43

D

Dadhich, Naresh K., 66, 67
Dain, Sergio, 43
Damour, Thibault, 43
Daniel, J., 43
Das, S.R., 43, 44
Das, Saurya, 21, 44
Das, Tapas K., 44
Dasgupta, A, 44
Davis, Amanda., 2
de Alwis, S.P., 45, 95
de Boer, Jan, 19
de Moura, Alessandro P. S., 45
Deane, J.R., 45
Degura, Yoshitaka, 45
Dehghani, M. H., 45
Demma, Th., 41
Denef, Frederik, 25
Dereli, Tekin, 45, 46
Diamandis, G. A., 38
Dias, Óscar J. C., 46
Dimopoulos, Savas, 46
Done, Chris, 112
Dove, J. B., 109
Dowker, F. , 46
Dreyer, Olaf, 46
Duff, M.J., 47
Dunning-Davies, J., 47
Dvali, Gia, 47
Dyer, A. (Alan), 2

Dzhunushaliev, V., 47

E

Easther, Richard, 47
Ellis, J., 48
Emparan, Roberto, 37, 48
Enderlein, J., 48
Eracleous, Michael, 96
Erlich, Joshua, 36
Etesi, Gá, 48

F

Fabbri, Alessandro, 41, 49
Fabian, A. C., 49, 65, 109
Fabian, A., 91
Fairhurst, Stephen, 16
Feigelson, E. D., 57
Fender, R.P., 49
Feng, Jonathan L., 50
Ferrara, Sergio, 37, 50
Ferrarese, Laura, 50
Ferrari, Valeria, 50, 97
Field, George, 51
Fineberg, J, 113
Firsova, N. E., 51
Fischler, W., 20
Fiziev, Plamen P., 110
Flambaum, V. V., 51
Ford, Kenneth William, 11
Fré, P., 2, 25, 51
Freiburghaus, C, 94
Frolov, V. P. (Valerii Pavlovich), 3, 51, 52
Fryer, Chris L., 52
Fuller, G.M., 98

G

Gabadadze, Gregory, 47
Gaida, Ingo, 53
Gal, Susan, 5
Galloway, G. J., 53
Gal'tsov, Dmitri V., 53
Gamboa, J., 54
Gao, Sijie, 54
Garay, L. J., 54
García, Alberto, 17
Garmire, G., 57
Garriga, Jaume, 54
Gauntlett, Jerome P., 54
Gegenberg, J., 44
Georgalas, B. C., 38

Ghosh, Amit, 44, 46
Ghosh, S. G., 55
Gibbons, G.W., 37, 55
Giddings, Steven B., 55, 56
Gilfanov, M., 38
Glampedakis, Kostas, 56
Gnedin, Oleg Y., 56
Gomberoff, Andrés, 20
Gómez, Roberto, 56
Gonçalves, Sérgio.M.C.V., 56
Goncharov, Yu P., 56
Goodman, Polly, 3
Gorini, V., 51
Gould, Andrew, 57
Gour, Gilad, 57
Govindarajan, T.R., 57
Green, Anne M., 57
Greene, B. (Brian), 8
Gregory, James P., 57
Gregory, Ruth, 48
Griffiths, R. E., 57
Grivell, Ian J., 75
Grumiller, D., 57
Gubser, S.S., 58
Guido, D., 58
Gurtug, Ozay, 58
Gutowski, J., 58

H

Haardt, F., 91
Haehnelt, Martin G., 58
Hagsted, Maj-Britt, 5
Haiman, Zoltan, 59
Halbersma, Rein, 22
Halzen, F., 59
Hambli, N., 59
Hammond, Richard T., 3
Hansen, R. N., 59
Haque-Copilah, S., 21
Harms, B., 35
Hartmann, Betti, 29, 60
Hawking, S. W. (Stephen W.), 3, 36, 60
Hawkins, Michael, 3, 4
Hayashida, K., 65
Hayward, Sean A., 60, 95
Hazen, Robert M., 4
Heckler, A.F., 61
Heindl, W. A., 109
Helfer, Adam D., 61
Hemming, Samuli, 61
Hemsendorf, Marc., 4
Heusler, M., 96

Higuchi, Atsushi, 41, 61
Hirayama, Takayuki, 61
Hirschmann, Eric W., 38
Hiscock, William A., 15, 103
Hjelmeland, S.E., 62
Ho, Jeongwon, 62
Hod, Shahar, 62
Holst, Sören, 15, 62
Hong, Soon-Tae, 62, 63
Horowitz, Gary T., 48, 63, 64
Hu, L, 78
Huang, Wung-Hong, 64
Hübner, Peter, 64
Hughes, Scott A., 64, 65
Huq, Mijan, 83
Husa, Sascha, 56, 65
Husain, V., 44

I

Iizuka, Norihiro, 65
Iwasawa, K., 19, 65

J

Jackson, Mark G., 65
Jacobson, Ted, 65
Jakubi, Alejandro.S., 38
Jalali, Mir Abbas, 66
Jaranowski, Piotr, 66
Jedamzik, K., 66
Jenet, F. A., 66
Jhingan, S., 66
Ji, Jeong-Young, 33
Jing, Jiliang, 67
Johnson, Clifford V., 37, 48, 67
Joshi, Pankaj S., 67

K

Kabat, Daniel, 65
Kallosh, R., 67, 68
Kan, Nahomi, 69
Kane, Gordon L., 14
Kanti, P., 69
Kaplan, D.M., 69
Kaplan, David B., 39
Kapusta, J. I., 69
Kar, Sayan, 69, 70
Karas, V., 83
Kastor, D, 46
Kastrup, H. A., 26
Kawaler, Steven D., 4

Kelly, Bernard, 70
Kerimo, Johannes, 70
Keski-Vakkuri, Esko, 19, 70
Ketov, S. V. (Sergei Vladimirovich), 4
Khanna, Gaurav, 70
Kidder, Lawrence E., 71
Kiefer, Claus, 29
Kim, Hee Il, 71
Kim, Hongsu, 71
Kim, N.J., 75
Kim, Nakwoo, 54
Kim, S.K. , 71
Kim, S.P. , 71
Kim, Won Tae, 62, 63
Kim, Yong-Wan, 62
Klebanov, I.R., 20, 71
Kleber, B, 113
Kleihaus, Burkhard, 60, 71
Klemm, D., 31, 32, 71
Klemm, Dietmar, 71
Knutt-Wehlau, M.E., 72
Kohri, K., 72
Kol, B., 72
Kolanović, Marko, 47
Krasnitz, M, 72
Krasnov, Kirill V., 72, 73
Kraus, P., 19
Krennrich, F., 73
Kribs, Graham D., 73
Kummer, W., 74
Kunz, Jutta, 29
Kupi, Gabor, 74

L

Laguna, Pablo, 70, 83
Lanfermann, Gerd, 14
Larsen, A.L., 74
Larsen, F., 42, 74, 75
Lasota, Jean-Pierre, 75
Lawrence, A., 19, 65
Le Bohec, S., 73
Leach, Samuel M., 75
Leblanc, Y., 35
Lee, Chul Hoon, 71, 75
Lee, H.W., 75
Lee, Hyun Kyu, 75
Lee, Hyung Mok, 76
Lee, William H., 76
Lehner, Luis, 83
Leibovich, Adam K., 73
Letelier, Patricio S., 15, 76
Lethem, Jonathan, 5

Levinson, Amir, 76, 77
Levy, David H., 10
Li, Miao, 77
Li, Zhong-heng, 77
Liang, Y. C., 77
Liberati, Stefano, 77
Lifschytz, Gilad, 65
Lin, Feng-Li, 78
Lin, L.-M., 28
Linet, B., 78
Lira, P., 65
Liu, Hongya, 78
Liu, Y, 78
Lockitch, Keith, 70
Longo, R., 58
Lousto, Carlos O., 78
Lowe, David A., 69, 79
Lu, Youjun, 79
Lubo, M., 79
Lue, Arthur, 79
Luomajoki, Markus, 80
Lyon, Carol, 8

M

MacCallum, M. A. H., 87
MacGibbon, J.H., 79
Macías, Alfredo, 17
Maciejewski, Witold, 80
Maeda, Kei-ichi, 103
Maeda, Kengo, 80
Magorrian, John, 80
Magueijo, João, 80
Mahapatra, Swapna, 53, 80
Mäkelä, Jarmo, 80
Maki, Takuya, 69
Makishima, Kazuo, 81
Maldacena, J., 81
Malzac, Julien, 81
Mandal, Gautam, 29
Manko, Vladimir S., 82
Mann, Robert, 82
Mao, Shude, 82
Marolf, Donald, 83
Marronetti, Pedro, 83
Marsh, T.R., 83
Martellini, M., 23
Martínez, Cristián, 83
Martinec, E., 77, 83
Martocchia, A., 83
Matarrese, S., 97
Mathur, Samir D., 43, 83
Matyjasek, J., 84

Matzner, Richard A., 83
Mavromatos, Nick E., 84
Mayo, Avraham E., 84
Mbonye, Manasse R., 14, 84
Medved, A. J. M., 84
Melia, F., 112
Mena, Filipe C., 85
Merloni, A., 91
Merritt, David, 85
Michaud, G., 29, 85
Michelson, J., 85
Mignemi, S., 85
Miller, J. Hillis (Joseph Hillis), 5
Miller, J.C., 36
Miller, Jake, 5
Miller, W.A., 29
Mishima, T., 89
Mitra, P., 85, 86
Mittaz, J. P. D., 89
Mohaupt, Thomas, 53, 86
Monaco, Pierluigi, 96
Morales, R., 86
Morawetz, Cathleen Synge, 86
Moretti, V., 71
Murali, Chigurupati, 86
Mushotzky, Richard F., 96
Myers, Janet Nuzum, 5
Myers, Robert C., 54
Myung, Yun Soo, 33, 75, 87

N

Natsuume, Makoto, 87
Navarro, Diego J., 41, 49
Nayak, K. Rajesh, 87
Nerger, Lars, 14
New, Kimberly.C.B., 87
Newton, David E., 6
Ng, Y. Jack, 87
Niemeyer, Jens C., 87
Nitta, Shin-ya, 88
Nojiri, Shin'ichi, 88, 105
Nollert, Hans-Peter, 88
Nomoto, Ken'ichi, 104
Novikov, I. D. (Igor Dmitrievich), 3, 4
Nowak, M. A., 109
Nucamendi, Ulises, 40, 41, 88

O

Obregon, Octavio, 88
Odintsov, Sergei D., 88
Okamura, T., 65

Olinto, A.V., 106
Olum, K.D., 89
Onozawa, H., 65
Opher, Reuven, 89
Ori, Amos, 89
Oxlade, Chris., 6

P

Page, Don N., 89
Page, M. J., 89
Pakis, Stathis, 54
Papadopoulos, Philippos, 90
Pariev, V.I., 29, 90
Parikh, M.K., 90
Park, Mu-In, 90
Parker, L., 27
Parker, Steve, 6
Paul, B. C., 90
Pauri, Massimo, 50
Pavlidou, Vasiliki, 90
Peet, Amanda W., 90
Peña, Igor, 91
Peterson, Bradley M., 91
Petrucci, P.O., 91
Pfeiffer, Harald P., 91
Phillips, Nicholas G., 91ú
Piasecki, M., 43
Pierre, J.M., 91
Pierro, V., 41
Pinto, I. M., 41
Plyatsko, Roman, 92
Podolsky, J., 92
Polarski, D., 92
Pollney, Denis, 14
Pomazanov, M. V., 14
Pons, Jose A., 92
Portegies Zwart, Simon F., 92
Pottschmidt, K., 109
Poutanen, Juri, 4
Prakash, Madappa, 92
Pravda, V., 92
Prestidge, Tim, 93
Pretorius, Frans, 93
Price, Richard H., 93
Priedhorsky, William C., 94
Proeyen, Antoine.Van, 25
Ptak, A., 57
Pullin, Jorge, 70
Punsly, Brian, 6

Q

Qian, Y, 93
Qiu, Weigang, 93

R

Rajaraman, A., 68
Ramón Medrano, M., 93
Rauscher, T, 94
Reall, Harvey S., 37
Reeder, Jesse, 7
Rees, Martin J., 93
Reid, D.D., 94
Reig, P., 94
Rembges, F, 94
Repo, Pasi, 80
Revnivtsev, Mikhail G., 94
Reznik, B., 94
Ridpath, Ian, 7
Roberts, J. E., 58
Rogatko, Marek, 95
Romine, J.S., 107
Rothman, Tony, 77
Rubinstein, Jonathan, 5
Ruiz, Eduardo, 82

S

Sachs, Ivo, 95
Sakai, Nobuyuki, 95
Salucci, Paolo, 96
Sambruna, R., 57
Sambruna, Rita M., 96
Santos, Caroline, 96
Sarbach, Olivier, 96
Sathyaprakash, B.S., 96
Sazhin, M. V., 14
Scardigli, Fabio, 97
Schatz, H, 94
Scheel, Mark A., 71
Schilling, Govert, 7
Schleich, Kristin, 53, 102
Schneider, R., 97
Schnetter, Erik, 70
Schramm, F., 26
Schutz, B.F., 97
Schutz, Bernard, 97
Schützhold, Ralf, 98
Segal, Justin., 7
Seidel, Edward, 14, 18
Semerák, O., 98
Sermutlu, Emre, 98

Shapiro, Stuart L., 99
Shi, X., 98
Shibata, Masaru, 98, 99
Shinkai, Hisa-aki, 37, 99
Shiromizu, Tetsuya, 99
Shmakova, M., 99
Shoemaker, Deirdre, 70
Shojai, Ali, 99
Sigurdsson, Steinn, 99
Singer, Maxine, 4
Singh, T. P., 99
Siopsis, George, 100
Sipe, J.C S., 107
Sipiera, Paul P., 8
Sivaram, C., 100
Slivka, L.S S., 107
Smalley, Larry L., 100
Smart, Brian, 9
Smoller, J. A., 101
Smoller, Joel, 101
Sobel, D.F S., 107
Sofue, Yoshiaki, 9
Soh, K., 71
Sonego, Sebastiano, 77
Sonoda, H., 101
Sood, Abha, 109
Sorkin, Evgeny, 101
Starkman, Glenn D., 101
Steiner, Andrew W., 92
Stevens, J. A., 89
Stone, James M., 102
Strathern, Paul, 8
Strominger, A., 69
Stryla, Jerzy, 102
Su, Ru-Keng, 93, 108
Surya, Sumati, 102
Suzuki, K., 102
Svensson, R. (Roland), 4
Szuszkiewicz, Ewa, 96, 102

T

Tamaki, Takashi, 102, 103
Tassis, Konstantinos, 90
Tavakol, Reza, 85
Taylor, Brett E., 103
Taylor, Edwin F., 9
Taylor, John Gerald, 9
Teitelboim, Claudio, 9, 83
Teukolsky, Saul A., 71, 91
Thielemann, F, 94
Tichy, Wolfgang, 104
Tomimatsu, Akira, 104

Torii, Takashi, 80
Townsley, L., 57
Troncoso, Ricardo, 16, 41
Tsuruta, Sachiko, 104
Turolla, Roberto, 112

U

Umeda, Hideyuki, 104

V

Valtancoli, P., 104
van der Bij, J. J., 105
van Nieuwenhuizen, Peter, 105
van Putten, Maurice H. P. M., 105
Varadarajan, Madhavan, 105
Vaz, Cenalo, 105, 106
Vendrell, F., 106
Venkatesan, A., 106
Verbunt, Frank, 106
Villas da Rocha, J. F., 28
Villiers, Jean-Pierre De, 106
Visser, Matt, 107
Vogel, P., 93
Vogt, Gregory, 11

W

Wadia, Spenta R., 107
Wagner, S S., 107
Wald, Robert M., 1
Wang, A. Z., 28
Wang, Bin, 93, 107, 108, 113
Wardzinski, Grzegorz, 108
Weinberg, Martin D., 108
Wessels, Ewald, 108
Wheeler, John Archibald, 9, 11
Whittaker, Bibby, 9
Wilman, R. J., 49, 109
Wilms, J., 109
Wilson, Colin, 11
Winstanley, Elizabeth, 109
Wirschins, Marion, 109
Wiseman, Alan G., 109
Wit, Bernard de, 34
Witt, D. M., 53
Witt, Hans J., 82
Witten, Edward, 110
Witten, Louis, 106
Wolfson, Richard, 12
Woosley, S. E., 52
Wrobel, J. M., 110

Wu, X. N., 110

X

Xu, Dian-Yan, 110
Xu, Jianjun, 93

Y

Yamaguchi, Masahide, 110
Yamazaki, Ryo, 110
Yazadjiev, Stoytcho S., 110
Yo, Hwei-Jang, 111
Youm, Donam, 111

Yu, Qingjuan, 111
Yuan, Feng, 111
Yusef-Zadeh, F., 112

Z

Zahn, Timothy, 12
Zampieri, Luca, 112
Zanelli, Jorge, 9
Zaslavskii, O.B., 112
Zdotycki, Piotr T., 112
Zeff, B W, 113
Zhang, Shuang Nan, 41
Zhu, Jiong-Ming, 113

SUBJECT INDEX

A

absorption, 58, 71, 72, 81
abundance, 94, 109
acceleration, 59, 76, 94
accretion (Astrophysics), 4, 11
accretion Disks, 30, 84, 90
action integral, 74, 85
algebra, 21, 48
amplitudes, 95
angular momentum, 42, 54, 58, 81
anisotropy, 80
astronomers, 5
astronomy, 2, 3, 6-8, 10, 11
astrophysics, 3, 6, 10
atlases, 10
attractors, 68

B

background radiation, 31, 80
backscattering, 94
bag model, 66
basic interactions, 71
bifurcation, 52
binary stars, 24
black holes (Astronomy), 1-9, 11, 12
bosons, 58
bound state, 68, 75, 89
boundary conditions, 20, 27, 30, 70, 89

C

celestial mechanics, 7
charge density, 29
charged particles, 62
charges, 13, 21, 30, 45, 72, 92, 95, 99
chiral symmetry, 72

compactification, 19, 22, 69
composite models, 22, 58, 63, 68, 95
confinement, 66, 110
conformal invariance, 19, 20, 42, 100
conservation laws, 69
coordinates, 101
corrections, 22, 35, 88, 93
correlation functions, 42
cosmic gamma sources, 84
cosmic neutrinos, 59
cosmic ray flux, 80
cosmological constant, 25, 28, 39
cosmology conferences, 6
cosmology, 3, 6, 10, 25, 28-30, 54, 62, 63, 65, 66, 80
coupling constants, 71
coupling, 20, 63, 71, 72, 95, 110
criticism, 5
cross sections, 42, 58, 71, 72
curiosities and wonders, 9, 12

D

dark matter (astronomy), 3, 6
defects, 13
density matrix, 89
density, 66, 89, 110
difference (psychology) in literature., 5
differential equations, 24, 43
differential geometry, 62, 69, 85
dimensions, 14, 29, 47
discoveries in science, 5
dispersion relations, 43
distribution, 30
double stars, 1
duality, 2, 8, 19, 29, 30, 41, 42, 85
dwarf stars, 14
dynamics, 35, 38, 42
dyons, 54, 80, 90

E

einstein field equations, 31, 101, 107
Einstein, Albert, 1879-1955, 12
Einstein-Maxwell Equations, 30, 81, 90
electric charges, 35
electromagnetic fields, 43
electromagnetism, 44
electron-positron interactions, 43
element abundance, 93, 94
elementary particles, 58
emission spectra, 42
emission, 41, 42, 44, 76, 81, 90
energy conservation, 69
energy levels, 40
energy spectra, 23, 24, 43
energy transfer, 43
energy, 15, 23, 24, 27, 40, 43, 60, 68, 69, 89
energy-momentum tensor, 13, 74
entropy, 13, 20-22, 29, 32, 34, 35, 37, 40-42, 52, 54, 59, 63, 69, 71, 78, 79, 81, 85, 87, 89, 90, 93, 95, 99, 102, 107, 108
equations, 24, 43, 74, 85, 90, 95, 101
evaporation, 49, 89
evaporation, 89
excitation, 42, 81
expectation value, 89
extended particle model, 22, 40, 46, 58, 68, 75, 95

F

fermions, 13, 58
Feynman Path Integral, 24, 90
field theories, 22, 24, 40, 43, 58, 71, 75, 95
fluctuations, 30, 52, 66, 72, 78
fluid mechanics, 90
fluids, 90
fluorescence, 30
four-dimensional calculations, 42, 58, 75, 85, 102
free energy, 27
functions, 24, 27, 42

G

galactic center, 9
galactic evolution, 13, 59
galactic nuclei, 1, 4
galaxies, 10, 13, 21, 50, 61, 110
gamma radiation, 61
gamma ray astronomy, 4
gamma ray bursts, 7
Gases, 74

general relativity theory, 25, 31, 34, 40, 43, 44, 48, 62, 71, 74, 75, 85, 107
geometry, 27, 42, 44, 63, 68, 99, 107, 108
global analysis, 25
gluons, 61
gravitation, 6, 19, 20, 24, 31, 44, 101
gravitational collapse, 28
gravitational interactions, 71
gravitational radiation, 23, 24, 40, 78
Gravitational Waves, 1, 24, 25
gravitons, 89
Great Britain, 7
green function, 23, 24
ground states, 40

H

hamiltonian function, 85
hamiltonians, 27, 47, 52, 90
handicraft, 3
Hawking, Steven, 3, 8
Higgs Model, 13
Hillis (Joseph Hillis), 5
humorous stories, 5
hydrodynamics, 38, 76
hydrogen Burning, 94
hydrogen, 84, 94

I

impact parameter, 44
inflation, 28, 30
instability, 15, 18, 54, 64
integrals, 24
interactions, 43, 71
iron, 52, 111

J

juvenile literature, 2, 6, 8

K

Kaluza-Klein Theory, 54
Kerr Metric, 48
Klein-Gordon Equation, 23, 24

L

lagrangian function, 74
lepton-lepton interactions, 43
light cone, 44, 79
line widths, 30, 90

locality, 89
Lorentz Invariance, 79
love stories, 5

M

magnetic fields, 38
magnetohydrodynamics, 6, 38, 43
many-dimensional calculations, 21, 37, 44, 68, 71, 81
mass transfer, 24
mass, 13, 21, 24, 25, 29, 41, 44, 66, 74, 96
massless particles, 31, 58
mathematical models, 22, 40, 58, 68, 75, 93, 95
mathematical operators, 27
mathematics, 27, 30, 31, 68, 89
matrices, 77, 89
matrix elements, 63, 79, 81
matter, 20, 37, 66, 69, 71
Maxwell Equations, 35, 43
mechanics, 16, 85, 101
membranes, 54, 74, 90
metrics, 19, 24, 27, 43
microwave radiation, 31, 80
milky way, 9, 10
mixing, 72
mode conversion, 43
monopoles, 13, 26, 56
motion, 74, 90
multiplets, 68

N

Navier-Stokes Equations, 90
neutrino detection, 98
neutrino oscillation, 78
neutrinos, 77, 78
neutron reactions, 84
neutron stars, 4, 9, 14, 24, 52, 84, 106
nonluminous matter, 66
nucleation, 28
nucleosynthesis, 66, 93
numerical analysis, 30, 31, 89, 101

O

occupational neuroses, 5
occupations, 5
ohm law, 90
opacity, 52
orbits, 23, 24
outer space, 2, 6

P

pair production, 13, 54
Partial Differential Equations, 24, 43, 80
particle interactions, 43
particle models, 22, 58, 68, 75, 95
particle properties, 58
partition functions, 27
Perturbation Theory, 31, 66, 89, 107
phase diagrams, 83
phase space, 85
phase transformations, 77
philosophy, 3
photoemission, 43
photons, 31, 58, 89
physical properties, 22, 27, 71, 89, 95
physicists, 11
physics, 1, 2, 4, 13-16, 18-24, 26-28, 31-34, 36-40, 44, 45, 47-64, 67, 70, 72-99, 101, 102, 104, 106-112
planets, 7, 10, 14
plasma astrophysics, 6
plasma instability, 38
plasma simulation, 43
plasma waves, 43
plasma, 38, 43, 81, 89
Poincare Groups, 85
populations, 14
potential scattering, 100
probability, 30, 52, 90
probes, 81
propagator, 23, 24
Proust, Marcel 5
pulsars, 24, 52

Q

quantization, 13, 23, 24, 41, 85, 94, 100
quantum chromodynamics, 101
Quantum Field Theory, 19, 20, 24, 40, 58, 71, 81, 95
Quantum Gravity, 6, 13, 15, 16, 18-26, 28, 30-36, 38-41, 44-47, 49, 51-58, 60-64, 67, 69-73, 75-78, 80-82, 84, 86-100, 102-104, 106-110, 112
quantum mechanics, 35, 40
quantum operators, 27
quantum theory, 6, 12
quark model, 22, 40, 58, 63, 68, 95
quarks, 61, 92
quasars, 21, 36, 98
quasi particles, 22, 68

R

R process, 93
radiations, 24, 40, 45, 72, 81, 94, 95
red shift, 98
relativistic astrophysics, 11
relativistic plasma, 38, 43
relativity (physics) space and time, 12
renormalization, 71, 81
reviews, 13, 38, 51, 58, 64
rotation, 42, 48, 68, 81

S

scalar fields, 23, 24, 31, 42, 44, 62, 70, 71, 72
scalars, 58, 69, 71, 72, 81
scaling laws, 33, 62
scaling, 14, 26, 33, 58
scattering amplitudes/eikonal approximation, 44
scattering, 36, 45, 54, 56, 85, 95
Schwarzschild Metric, 23, 24, 25, 27, 38, 43, 44, 63
Schwarzschild Radius, 63
schwinger functional equations, 101
science, 12
selection rules, 13
semiclassical approximation, 45, 58, 81, 95, 102, 107
seyfert galaxies, 30, 90
sigma particles, 5
simulation, 30, 43
singularities (mathematics), 4
singularity, 27, 31, 44, 54, 63, 74, 89, 101, 113
size, 14
solar neutrinos, 78
solar system, 2, 3, 7, 10
solitons, 22, 29, 39, 41, 44, 63, 68, 81, 85, 92
sorption, 58
space and time, 1, 4
Space sciences, 6
space, 38, 74, 103, 110
space-time, 19, 25, 27, 31, 33, 42, 46, 52, 62, 63, 68, 69, 70, 74, 80, 85, 87, 92
sparticles, 89
spectra, 24, 30, 44, 61, 66, 90, 94
spectroscopy, 43
spin, 58
stability, 38, 60, 98
star accretion, 24, 52
star evolution, 13, 24, 52
star models, 98
stars, 7, 11, 14, 24, 80, 92, 94
statistical mechanics, 69, 74
statistics, 24
stellar atmospheres, 52
stellar dynamics, 4
stellar radiation, 42
stellar winds, 52
stochastic processes, 30
stresses, 69
string models, 102
string models, 2, 8, 13, 19, 21, 22, 42, 44, 46, 58, 68, 69, 71, 75, 77, 79-81, 92, 95
Su Groups, 101
Su-2 Groups, 101
supergravity, 22, 29, 58, 63, 68, 69, 72, 77, 81, 85, 89, 99
Supernova Remnants, 52
Supernovae, 6, 24, 52, 93
superstring models, 22, 29, 69
superstring theories, 8, 9
supersymmetry, 5, 19, 21, 29, 37, 40, 41, 58, 63, 68, 69, 77, 79, 81, 83, 85, 89, 99
symmetry, 40, 58, 68

T

T Invariance, 29
tensors, 19
thermodynamic properties, 22, 27, 42, 71, 89, 95
thermodynamics, 14, 27, 33, 35, 46, 63, 69, 71, 74, 77, 79, 83, 88, 90, 92, 100
topology, 13, 27
torsion, 85
transformations, 89
transition amplitudes, 94, 95
travelling waves, 63
Trollope, Anthony, 5
tunneling, 54

U

ultraviolet divergences, 52
Unified-Field Theories, 43, 58, 63, 83
United States, 1
universe, 10, 11, 28, 37, 45, 54, 69, 97, 104

V

vacuum states, 22, 37, 62, 89
variable stars, 94
variations, 30, 86
vectors, 68
violations, 78
vocational guidance, 5
vortices, 13, 96

W

wave equations, 24, 42, 43
wave functions, 42
Wheeler, John Archibald, 9, 11
White Dwarf Stars, 14, 84
white dwarfs, 4
WKB Approximation, 54

X

X-ray astronomy, 4
X-ray background (XRB), 49, 96, 109

Y

Yang-Mills Theory, 63, 77, 83